Frontiers of Modern Physics

New Perspectives on Cosmology,
Relativity, Black Holes
and Extraterrestrial Intelligence

TONY ROTHMAN

and Bernard Carr, Richard Matzner, A. C. Ottewill,
Tsvi Piran, L. C. Shepley, E. C. G. Sudarshan,
Frank J. Tipler, Bill Unruh

DOVER PUBLICATIONS, INC.
NEW YORK

Published in Canada by General Publishing Company, Ltd., 30 Lesmill Road, Don Mills, Toronto, Ontario.
Published in the United Kingdom by Constable and Company, Ltd., 10 Orange Street, London WC2H 7EG.

Frontiers of Modern Physics: New Perspectives on Cosmology, Relativity, Black Holes and Extraterrestrial Intelligence is a new work, first published by Dover Publications, Inc., in 1985. Five of the seven articles that appear in this book have been published previously, either in part or in whole. "On Cosmology," by Tony Rothman and L. C. Shepley, appeared in the July 1979 issue of *Isaac Asimov's Science Fiction Magazine.* "Demythologizing the Black Hole," by Richard Matzner, Tsvi Piran and Tony Rothman, was published in the September 1980 issue of *Analog.* "The New Neutrinos," by Richard Matzner and Tony Rothman, appeared previously in the May 25, 1981 issue of *Analog.* "Coincidences in Nature and the Hunt for the Anthropic Principle," by Bernard Carr and Tony Rothman, appeared in a shorter version in the October 26, 1981 issue of *Isaac Asimov's Science Fiction Magazine.* Parts of "Extraterrestrial Intelligent Beings Do Not Exist," by Frank J. Tipler, appeared in three issues of *The Quarterly Journal of the Royal Astronomical Society* in 1980 and 1981. "The Peryton and the Ants," by E. C. G. Sudarshan and Tony Rothman, and "Grand Illusions: Further Conversations on the Edge of Spacetime," by Richard Matzner, Tony Rothman and Bill Unruh, appear for the first time in this book. The appendix to "Grand Illusions"—"Communication with Observers Falling into Black Holes"—by A. C. Ottewill and Tony Rothman, also appears for the first time in this book.

Manufactured in the United States of America
Dover Publications, Inc., 31 East 2nd Street, Mineola, N.Y. 11501

Library of Congress Cataloging in Publication Data

Main entry under title:

Frontiers of modern physics.

1. Cosmology.   2. Relativity (Physics)   3. Black holes (Astronomy)   4. Life on other planets. I. Rothman, Tony.
QB981.F76   1985   523.1   84-13623
ISBN 0-486-24587-X

Contents

Preface

Recently we have witnessed a well-documented boom in the public's interest in science. Relativity captured the physical science spotlight during the Einstein centennial in 1979. Computer sciences, virtually unknown twenty-five years ago, now have entire magazines devoted to them. The possibility of life-extending and memory-enhancing drugs captures our imagination. For the most part, this interest cannot be bad, but it strikes some of us as misdirected. In the new popular magazines devoted exclusively to science, where the average article length does not exceed one thousand words and in which the latest discoveries are encapsulated in fifty-word bits, the most important aspect of science is missing altogether: process.

The most important aspect of painting is the *painting*, not the portrait. The most important thing in writing is the *writing*, not the novel. Music is so vital because—perhaps more than any other art form—it allows musician and audience to experience *process* with each performance. So it is with science. The discovery is not as important as the search.

In these articles, therefore, you will hear surprisingly little of, "Scientists have concluded the universe was created 15.7 billion years ago in an explosion isotropic to one part in . . . ." Our statements will often end with question marks. You will be subjected to the reasoning process, insofar as it can be presented, without elaborate mathematics. Sometimes the arguments are long and difficult in concept. You will be required to think. You will not always understand everything. You will also occasionally be subjected to opinions of ours that grate on your nerves. This is okay. Contrary to popular mythology, scientists are people too, and they often hold annoying opinions.

Although the articles in this collection were written at different times and by different authors, you will notice common threads running through several works. There is a concern with models—those constructs within whose framework we view the world. There is a concern about how our own perception plays a role in the physical theories we develop. The role of perception in physics is becoming of greater and greater concern to physicists. The first halting steps are being taken to incorporate the experimenter into quantum mechanics and relativity. At the moment the steps are more philosophical than physical and some will argue it can never be anything else. Time will tell.

We cannot claim that all the articles were written with such high philosophical intent. One or two were commissioned; others were composed just for fun. Most often we saw an area in which the public was lacking information or was just plain misinformed, and we self-righteously took it upon ourselves to correct the situation. This approach can cause problems. One editor wrote in rejection: "You have explained why a certain mechanism won't work which most people have never heard of in the first place."

Communication is difficult. So is collaboration. Nonetheless, on the presumption that no single person knows everything and that scientists know their fields better than journalists, the attempt was made. The physicists here can legitimately be called leading experts. Most are either permanent staff or have been visitors at the University of Texas Center for Relativity Theory in Austin, the largest establishment for the study of the subject in the United States. Therefore, most of the articles touch on relativity. By and large, all the articles are true collaborations. It cannot be hoped that the articles are uniformly successful, but it is hoped that you will find in them new opinions with which to argue and new areas of thought to explore.

TONY ROTHMAN, in collaboration

Tony Rothman received a B.A. in physics from Swarthmore College in 1975 and a PhD from the Center for Relativity at the University of Texas, Austin, in 1981. Since then he has done post-doctoral work at the Department of Astrophysics, Oxford University, and at the Shternberg Astronomical Institute, Moscow, and is spending 1984–1986 at the University of Cape Town.

He is also the author of a novel, The World Is Round *(Ballantine/ Del Rey, 1978), and a play,* The Magician and the Fool, *which won the Oxford Experimental Theatre Club Competition for 1981–1982. His shorter works have appeared in various magazines and he is currently completing a large novel,* Apocrypha. *When time permits he plays the oboe and studies Russian.*

Frontiers of
Modern Physics

1

On Cosmology

by Tony Rothman & L. C. Shepley

At the time I began this series of articles, popular expositions of relativity and cosmology had already started to appear on the market. The authors of such treatments concentrated uniformly on what cosmologists term the "standard model" of the universe, and the public may have been left with the impression that there was nothing else to the subject. Deciding to remedy the situation, I enlisted the help of L. C. Shepley, who is an expert in nonstandard cosmological models. He graduated from Swarthmore College in 1961 and received his PhD from Princeton in 1965. Professor Shepley is now Associate Director of the Center for Relativity at the University of Texas in Austin and is co-author with Mike Ryan of the text Homogeneous Relativistic Cosmologies (Princeton University Press). In addition to his cosmological pursuits, L. C. is a gourmet chef, plays the flute and organ and is learning Spanish.

The original article was written as a dialogue between the two of us in a well-known Austin beer garden. It was also meant to be a tale from the 1001 Nights. The editor rejected this approach. Moreover, we assumed the reader had heard of the standard model. In retrospect, this was probably a mistake. At the risk of some redundancy, I have therefore added for this edition a short note on the standard model and have made a few corrections in the body of the text. Otherwise, the article is substantially as it appeared in the July 1979 issue of Isaac Asimov's Science Fiction Magazine. I do regret the title; for years after the article's publication, all science articles in IASFM have begun "On . . . ."

Note on the Standard Model

The "standard cosmological model" assumes that the universe started off at some time in the past in a "big bang." Although the theory itself does not say when this big bang occurred, direct astronomical evidence indicates that it was between 10 and 20 billion years ago. According to the standard model, only radiation and neutrons and protons existed one second after the big bang. The universe was much too hot for heavier elements and isotopes to stick together. As the universe cooled to about 1 billion degrees, however, the neutrons and protons collided to form deuterium and, from that point on, deuterium was burned into tritium and helium. This burning started at almost precisely three minutes after the big bang. After about a day the universe became too cold for any nuclear reactions to proceed, and the burning ceased. One of the standard model's greatest successes is the prediction that, after the completion of nuclear burning, the universe should contain about 25% helium by mass and 2×10^{-5} parts deuterium—roughly what is actually observed. (For more details, you are referred to Steven Weinberg's excellent book, *The First Three Minutes*.) The other great success of the standard model is the prediction that the radiation left over from the initial explosion should be visible today. In 1965 this radiation was indeed discovered and is now known as the "cosmic background radiation."

As discussed in this article, the standard model has some problems, however. At the instant of the big bang itself, all physical quantities—pressure, temperature, curvature—become infinite, and the theory breaks down. In addition, the standard model assumes that the universe is a uniform volume of gas expanding at equal rates in all directions (isotropic expansion). If the universe is so even, it is difficult to see how galaxies, which are manifest fluctuations on the even background, could have arisen (see, however, the postscript to "The New Neutrinos" in this volume). It is for these reasons and others that cosmologists study nonstandard cosmological models, the subject of this article.

Relativity theory is currently receiving a good press. The public, as well as we relativists, have been captivated by the charms of quasars, pulsars, black holes and expanding and collapsing universes. The excitement dispersed by our popularizers would have the layman believe that we have succeeded in sewing up the universe—with perhaps a minor loose thread here and there.

The outlook from the center of the known universe—that is, from the Center for Relativity Theory in Austin, Texas—is slightly less complacent, although not less enthusiastic. Nonspecialists are often unaware that cosmology theory is currently struggling with a dilemma. The big bang model of the universe, which has become known as "the standard model," explains the general features of the universe fairly well. However, it fails to explain several details—galaxy formation, for example—and has the fundamental problem that the big bang itself is a physical impossibility. Alternative models that try to be fully consistent and to explain several features that the standard model leaves out fail on both counts; that is, they are not self-consistent and do not explain the real world any more fully. Much of the work in current cosmology theory is devoted to exploring nonstandard models that often produce features so bizarre that one may be left wondering if current-day physics has anything to do with reality.

We would like to take you one step beyond the well-known big bang model and explore some of these nonstandard models. One of us believes this exploration to be physically relevant; the other has not yet made up his mind. We leave it to you to decide whether we and other relativists have gone off the deep end.

To start the discussion, consider cosmology, the study of the structure of the universe. All experimental evidence to date indicates that the universe is highly ISOTROPIC: *The visible universe has the same properties in all directions.* (The reader is advised to memorize the preceding sentence.) Antennas receiving 1-centimeter-wavelength radiation from space record the same amount of flux no matter which direction they point to. Flux is defined as the amount of energy from a source received by a square centimeter of antenna during one second. If you think of grass as an antenna, it receives a certain amount of energy from the sun over each square centimeter each second. This is the solar flux. So it is with the universe: All of space is permeated with radiation—the so-called cosmic black body radiation. It is now known that this radiation must come from the very early stages of the universe, just 100,000 years after the apparent time of the big bang. The flux of this radiation is extraordinarily isotropic; it is exactly the same in all directions, to one part in 10^4.

The reader might ask whether the fact that galaxies are irregularly lumped in clusters means that the universe is not isotropic. After all, the earth does not have the same physical properties in all directions (mountains here, valleys there, oceans everywhere), and is therefore nonisotropic. The problem here lies in not looking on a large enough scale. On a scale the size of the universe, the clustering of galaxies disappears and matter is evenly spread, just as the earth looks like a smooth ball when viewed from a large enough distance. Direct astronomical observation shows this isotropy but there is also indirect evidence: The abundance of the element helium is less than can be produced in a nonisotropic universe. Yes, on the large scale, the universe is strikingly regular, even monotonous. This isotropy is the reason we have adopted as the "standard" model of cosmology an isotropic volume of gas expanding since the big bang 10^{10} years ago, gradually cooling and condensing into galaxies, stars and us.

The most common conception of cosmology is that it is the detailed study of physics within a realistic model of the universe. Most people conceive this model to be the standard model. However, as we have mentioned, relativists spend ninety percent of their time studying imagined universes that are clearly nonrealistic. We are accustomed to viewing physics, including cosmology, as the study of a real world, an investigation of the laws of nature. A legitimate question to ask is why we study such unnatural cosmologies as the causality-breaking Taub-NUT model, the empty Kasner model, the nonexpanding Gödel model and the nine Bianchi classes. Many of these models aren't even approximately isotropic and have other strange properties as well. At first glance, these models seem to bear absolutely no relationship to nature. After all, we don't see causality-breaking in the real world . . . the universe is not empty . . . the universe is expanding . . . the universe is highly isotropic.

A legitimate answer to this question might be the following: The above models and other nonisotropic models (or *anisotropic* models, as they are called in the trade) are investigated not only because the universe may have been anisotropic at one time, but also because it is important to know how the universe is not constructed in order to understand the way it *is* constructed. Even the "standard" isotropic big bang model discussed earlier was originally rejected by Einstein because it portrayed the universe as expanding. Einstein's biggest mistake was his assumption that he knew *a priori* what the universe was like—that it was static. If Einstein had investigated expanding models, the world would have been better prepared for Hubble's discovery that the universe was expanding.

A question still remains, however, concerning what are legitimate models for study. A scientific model is a construct, an application of a general physical law to a specific circumstance. We take the laws as initial assumptions. Then, perhaps, we add some additional constraints—the assumption of isotropy in cosmology, for instance. We construct a model based on these assumptions and compare it with reality—assuming our conception of reality means anything. This comparison is the way a scientific theory is tested, and the theory stands or falls on the results of the tests. The question is how many features of the "real world" should a model attempt to mimic. What, for instance, does a model of an empty universe such as the Taub-NUT model (which we will be using as an example) tell us about the way the real universe is or is not constructed? The real universe is, after all, self-evidently not empty. What does an empty model tell us that isn't obviously nonsensical?

In answering this question, we first make the observation that the real universe *is* empty, or almost empty. The average matter density of the universe is about 10^{-30} grams per cubic centimeter, the equivalent of 1 hydrogen atom per cubic meter. Yet, in an isotropic model that is supposedly patterned after the real world, it is only the effect of the matter which governs the expansion. (That is, the gravity caused by the matter in the universe is the only thing that determines how fast the universe expands.) This tiny fraction of matter in otherwise empty space is responsible for the behavior of the entire model back to the very instant of the big bang itself. This sole dependence on matter would not necessarily be bad if the model behaved properly, but unfortunately it doesn't. The starting point of the model, the big bang, is supposedly an infinitely dense event, a point at which all the matter in the universe is located. This infinitely dense state is a singularity of the model, a breakdown in the sense that the model can't describe reality before that point. We would like to remedy this situation. To show how a totally empty universe fits in, we must first look at anisotropic models in general.

Anisotropic models were originally studied in the hope that the singularity just spoken of might be avoided. To understand this hope, see the two illustrations in Fig. 1, which represent the universe as a cloud of dust particles collapsing under the influence of its own gravity. Now, particles that collapse perfectly isotropically, as in Fig. 1a, will meet at an infinitely dense point from which they cannot emerge. On the other hand, even a slight amount of anisotropy should allow the particles to miss each other. The picture can, therefore, be continued to describe the reexpansion (see Fig. 1b). This "near miss and reexpansion" is one way of viewing the big bang: an expansion following a previous collapse, or just a collapse in reverse.

As mentioned, the entire behavior of the standard model is governed by its matter and radiation content. In anisotropic models, the geometry, or "shape," of spacetime itself helps determine the evolu-

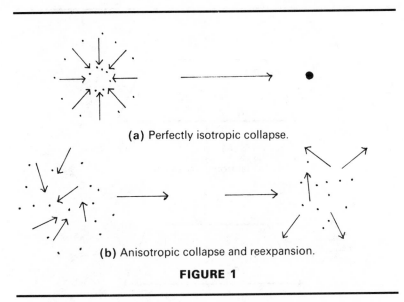

(a) Perfectly isotropic collapse.

(b) Anisotropic collapse and reexpansion.

FIGURE 1

tion of the universe. (These terms will be explained shortly, don't worry.) In fact, in a model which has matter and is even slightly anisotropic, the effect of pure geometry eventually dominates as we go back to the starting time of the model. The following model may portray the universe very well: Matter is most important when the universe is expanded and isotropic; geometry is most important near the big bang, if the universe turns out to be anisotropic there. Thus an empty model, one that is pure geometry, might be a good way to describe the early universe. Actually, the hopes for a nonsingular model proved futile; even an anisotropic model has a singularity, a point which will be explained later.

At this moment, you are confused and probably screaming, "What is all this nonsense about 'pure geometry'? Do you mean 'pure vacuum'?" No, no, not vacuum. In order to begin to get some understanding about our use of the term "pure geometry," refer to the two illustrations in Fig. 2. In Fig. 2a we've drawn a universe expanding isotropically in all directions. As this "rubber sheet" model of the universe expands, the dots, which on the average are smoothly spread, get farther apart. In Fig. 2b we've drawn little lumps of curved geometry. They are similar to lumps of matter in that, like matter, they have gravitational fields and therefore affect the overall rate of expansion. We will go into details below, but the important point to keep in mind for now is that in general relativity such curvature of

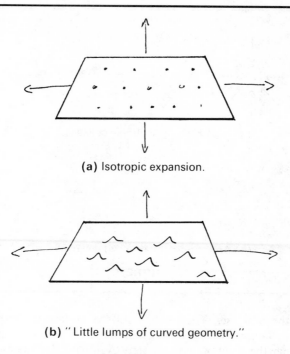

(a) Isotropic expansion.

(b) "Little lumps of curved geometry."

FIGURE 2

geometry is the source of a gravitational field in exactly the same way that matter is the source of a gravitational field. This is the explanation of why anisotropic models, unfortunately, possess a singularity. Consider again Fig. 1 with its collapsing cloud of particles representing the universe. The collapse drawn in Fig. 1b is anisotropic and thus has associated with it "little lumps of curved geometry." These lumps, as just mentioned, produce their own gravity, more than would exist in the isotropic case. Thus an anisotropic universe is actually pulled into a singularity faster than it would be if the anisotropy were absent. Thinking in reverse, it is then impossible to continue the reexpansion, or big bang, as we originally drew in Fig. 1b.

Now it is time to make these notions a bit more precise. For those of you who have heard that the presence of matter curves space, it might come as a surprise to know that space can be curved without matter. More properly, it is *spacetime* that is curved. Spacetime is the collection of all physical locations where events can occur. Since each event must be specified by its three coordinates of position and by its time, spacetime is four-dimensional. Spacetime has replaced the outmoded concept of two separate entities, "space" and "time," because they are now known to be inseparable. To see how spacetime can be curved without matter, we first examine the ordinary matter-induced curvature of spacetime.

Matter, by its presence, curves spacetime and therefore affects the local geometry. Other bodies, such as spaceships, attempt to move along the straightest possible paths, but these paths are themselves affected by the spacetime geometry produced by the first piece of matter and by all the other matter in the universe (see Fig. 3). In the absence of matter, spacetime is flat and particles move in straight

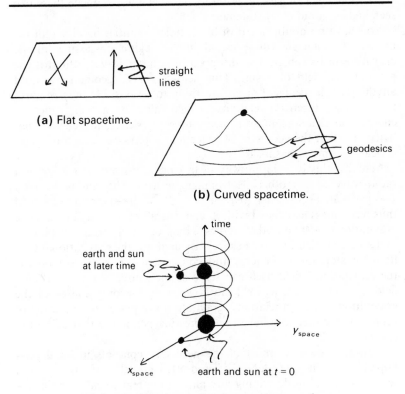

(a) Flat spacetime.

(b) Curved spacetime.

(c) Sun and Earth traveling forward in spacetime.

FIGURE 3

lines; that is, they move forward linearly in time and in space. Where matter causes curvature in the geometry, a particle moves along what is called a geodesic, the straightest path allowed. Think of the sun causing curvature, changing a flat geometry to a curved one. The earth moves around the sun in a spatial ellipse which, when combined with its motion forward in time, traces out a helix in spacetime. The helix is the straightest possible path allowed by the curvature. Farther away from the sun, Mars traces a more stretched-out helix because the effect of the sun on the curvature is less.

The bothersome question is: How can spacetime be curved without the sun or the presence of other matter? An empty spacetime should be flat; yet we've been talking about curved spacetime in the Taub-NUT model, which we said is empty. How can we reconcile these two seemingly contradictory statements?

You have no doubt heard of black holes—entities in which all the matter of a star has disappeared into a region possessing infinite curvature in its center. The disappearing matter leaves behind it the gravitational field of a star. This field is curved geometry without anything at the center. For such a thing to be possible only requires that we accept general relativity at face value. By $E = mc^2$, there is energy and therefore mass in everything physical, and general relativity says that the geometry of spacetime is physical. The conclusion is clear: Geometry has mass.

Actually, this is a sloppy way of talking. Consider the energy of a radio wave or any other electromagnetic field. The famous formula $E = mc^2$ can be inverted to read $m = E/c^2$. The energy of the field thus has a mass associated with it, and, therefore, like any mass, has a gravitational effect on other bodies. Think of the electromagnetic field as a sponge which can store energy. By analogy, the gravitational field itself is a storehouse of energy and, consequently, of mass. So you see, the gravitational field itself can serve as the source of a gravitational field. In some cases, as in Taub-NUT, it is the only source of the gravitational field. In relativity, gravitation is replaced by curvature of geometry; we therefore have the interesting possibility that curvature can cause itself.

Now that we have most of the necessary concepts in hand (and hopefully in mind), let's take the Taub-NUT model as an example and see if it is useful in describing any part of the real world. Part of this model was invented in 1951 by Abraham Taub. Newman, Unti, and Tamburino found an apparently unrelated model in 1963. It was Charles Misner who coined the notation NUT and showed that both are part of the same cosmological model. The Taub-NUT model is a case we have already mentioned: an empty universe in which curva-

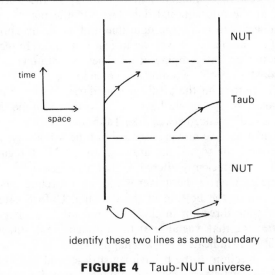

identify these two lines as same boundary

FIGURE 4 Taub-NUT universe.

ture exists without matter. In order to examine the properties that are caused by this absence of matter, we refer you to Fig. 4.

You will notice that the Taub-NUT model has three regions: two NUT regions separated by the Taub region. None of these regions exactly represents the real universe, although the Taub region is the closer of the two types. The general relativity equations that were used in deriving this model show that in the absence of matter the model must be highly anisotropic. Even the Taub region, therefore, differs from the real universe—first, in not having any matter, and second, in being anisotropic. The NUT regions, in some sense, look like the gravitational field of a black hole and originally were thought to be a new type of black hole model. Basically, the regions differ from each other in their geometry. This geometry is hard to describe without four-dimensional formulae, but we can draw some of the effects of the geometry in the various regions, as in Fig. 4.

You will see that we've drawn two vertical lines. These two lines are to be thought of as actually the same boundary for the following reason: The Taub region is closed in the same sense that the surface of a sphere is closed. In order to draw a map of the surface of a sphere on a flat piece of paper, we have to draw in an artificial boundary. When a particle traveling east reaches a boundary, it continues to travel east in reality, but on the map it looks as if it had suddenly jumped to the western boundary. Think of a map of the world with the boundary at the international date line. On the map a traveler leaving China seems to reach the date line and suddenly appear on the left side of the map as he travels on to San Francisco. In reality, of course, he doesn't do any jumping at all. The discontinuity is forced on us by the constraints of a two-dimensional map.

Although, by putting in the artificial boundaries, we've drawn the map to be closed in space, we have not closed it in time. This is because there is a difference between the Taub and NUT portions. In the Taub region, a particle moving forward in time will always appear to be going toward the top of the drawing. In the NUT region, the direction of time is sufficiently altered so that a particle which moves forward in time may circle the universe and, after circling, find itself moving back toward the bottom of the drawing. Unfortunately, we can't draw the time direction in the NUT portion accurately since we'd need a drawing that was at least three-dimensional. But we can at least schematically indicate what will happen (see Fig. 5).

The obvious objection to the behavior just described is that it violates causality. A particle started out moving forward in time and ended up moving backward in time—it could, therefore, conceivably enter its own past. Such behavior is usually enough to relegate any model immediately to the white elephant category.

But let us not jump to conclusions before we try to salvage some useful information from Taub-NUT. Remember that we had a particle

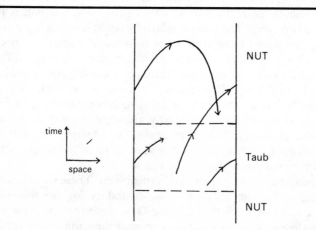

FIGURE 5 Motion of particles in Taub-NUT universe.

moving through spacetime even though we had previously stated the model was empty. This particle is known as a test particle. A test particle in physics is infinitely small and thus does not possess properties which disturb the quantities it is measuring. In the case of Taub-NUT, unfortunately, even the smallest amount of matter, even a test particle, destroys the model completely. The Taub portion is relatively unaffected, but the Taub-NUT boundaries are changed into singularities, leaving a Taub-like spacetime without the NUT attachments.

The reason that matter destroys the Taub-NUT boundaries is that some of the test particles circle the empty Taub region an infinite number of times before coming to the boundary. In the circling process a piece of matter comes arbitrarily close to the speed of light. It also comes arbitrarily close to itself, since it is circling like a cat chasing its tail. All of this is possible for two reasons. First, the circumference of the universe is becoming smaller and smaller near the boundaries, so it becomes possible for the particle to circle the universe in an arbitrarily short amount of time. Second, as the particle approaches the speed of light, its personal time ("proper" time) slows down, and it thus measures the time of circumnavigation going to zero.

Because the particle is coming infinitely close to itself and to any other test particles that may be present, the effect is that of an infinite density producing a singularity. Here, then, is a potentially useful piece of information from Taub-NUT. It is possible that the structure of the early, dense phases of the real universe, when the circumference was very small, involved such a process—namely, matter interacting with itself and with other matter from around the universe. In the standard model, however, near the time of the big bang a particle can interact with other matter only within a vanishingly small volume around itself.

This noninteraction among particles points to a big problem with the big bang model. If particles in the standard model cannot interact with each other near the big bang, then how does the universe "know" how to start off so regularly and isotropically? Since it can't have any interaction with the others, why doesn't each particle behave in any way it chooses, thereby producing a chaotic result instead of the isotropic universe we observe today?

This problem with the standard model is another argument for looking at offbeat models. For example, in trying to answer the above question, Charles Misner described just such a chaotic universe, in which particles behave in any way they please but which nonetheless becomes regular as it expands. He called it the Mixmaster model.

Misner patterned his Mixmaster model after none other than the

Taub-NUT model. He discovered that in a large class of models, an initial anisotropy is damped out as the universe expands because the matter is homogenized. This homogenization involves an averaging or mixing process that blends the properties of one part of the universe with those from other parts—a process requiring particles from over large distances. These particles are envisioned as traveling around the universe in the same sense that particles travel around the Taub-NUT model. The Taub-NUT model is used as a starting point, and the Mixmaster process blends that initial anisotropy into isotropic mush.

By now you probably suspect that there is a bug somewhere in the Mixmaster model and that cosmic blender hypotheses do not work. And this is, in fact, the case. We know that the universe is isotropic now. We are just trying to show here that isotropy is consistent with anisotropy in the early universe. Since Misner published his model, the work of Barrow and others has shown that the abundances of deuterium and helium are indirect evidence that the universe was isotropic at very early times, about 200 seconds after the big bang. The Mixmaster model would probably still be anisotropic at that time; thus we are forced to continue our search for other models.

Before searching, however, let's look more closely at the details of the helium–deuterium abundance argument. The amount of these isotopes produced in the big bang depends largely on how fast the universe was expanding at the time. If anisotropy is added to the standard model, the universe expands more rapidly. (Recall, anisotropy caused the universe to collapse faster into a singularity. In the reverse situation, it causes the universe to expand more rapidly.) A rapidly expanding universe raises the amount of helium produced during nucleosynthesis. By requiring that the helium produced during the big bang be less than the observed limit of 25% by mass, Barrow showed that the universe must have been essentially isotropic during primordial nucleosynthesis. We can, by other tricks, lower the helium produced, but doing so always raises the deuterium above observed limits. Consequently, we are boxed in by the isotopes and are forced to conclude the universe was isotropic as early as 3 minutes after zero.

It should be mentioned that some scientists—Steven Weinberg, for example—find this line of reasoning unconvincing and are willing to consider that the amount of deuterium observed in the universe is a result of current astrophysical phenomena such as supernovae and quasars. Incidentally, one of the conditions necessary to produce the correct amount of helium and deuterium in the early universe is that there exists too little matter to halt the universe's expansion and it will therefore go on expanding forever.

Bothered by the possibility of eternal expansion, a number of

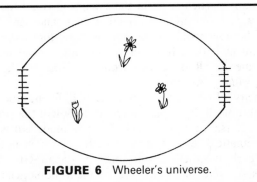

FIGURE 6 Wheeler's universe.

leading cosmologists have concentrated on models that recollapse—this recollapse going hand-in-hand with the closure of space. Working from the ideas of Einstein, John Wheeler for a long time concentrated on such models. In fact, he drew Fig. 6 for one of our students. John described the universe as a field of flowers with a gate at either end. Travelers don't know whether a gate is opened or closed until they get to it. The student was impressed.

We have by no means exhausted the various types of cosmological models; in fact, we have barely begun. The list could go on literally to infinity. To make life easier for the cosmologist, most of the models in this potentially infinite list are grouped into classes that describe their properties. These classes are based on mathematical rather than on physical ideas. This immediately creates the problem that most of the models within the classes are suspect as far as their "reality coefficient"* goes. Whether these models have a low enough reality coefficient to justify throwing them out is one question; what the rationale is behind the mathematical classification scheme is another.

First, we will talk about the mathematical classification scheme. The most important—or at least the most studied—general classes of models are HOMOGENEOUS: *At a given time the universe looks the same at any point.* (Memorize this sentence also.) Contrast homogeneity with isotropy. Isotropy means that at a given point the universe looks the same in all directions. Homogeneity can exist without isotropy (but not vice versa). Consider a rubber sheet that lengthens faster than it widens. The expansion process is identical at every point on the

---

* The Rothman reality coefficient is a number, lying between 0 and 1, which is assigned to the statement under question. True statements receive a 1, obvious nonsense gets a 0.

sheet, but at any particular point the expansion is *not* the same in all directions—it is faster in one direction than another. Thus, in an anisotropic, homogeneous model, the universe looks the same from any one point as it does from any other, but the universe does not look the same in all directions from any one of these points—*homogeneity without isotropy.*

Homogeneity allows for a vast number of different models. Some of these, though not all, have phases remarkably like the present epoch of the real universe. In order to classify these models, it is convenient to use mathematical ideas, the most useful of which is the work of Bianchi, an early twentieth-century mathematician. His work in pure mathematics showed that space can be made homogeneous in nine ways. The nine classes of homogeneous cosmological models are called the nine Bianchi types. Types I, V and IX are the ones most useful in such programs as Misner's attempt to show that a chaotic model automatically becomes isotropic as the universe expands. The Mixmaster model is a Type IX cosmology which is not initially isotropic, but later becomes so.

If you take all the potential models that the Bianchi types allow, you can come up with an infinite number of models. And yet, we still have not solved our original problem: Why is the real universe extraordinarily isotropic? Let us reemphasize that the helium and deuterium abundances show that the universe must have been isotropic only 200 seconds after the big bang. None of the offbeat models we have been discussing can satisfy this condition and at the same time look like our universe in its present form.

There is a conceivable use for the Bianchi types, nonetheless, however fantastical it may sound. This is in quantum cosmology. In quantum mechanics matter has wavelike properties. At a given point in spacetime the net effect of, say, a particle, is due not only to where the particle is, but to its entire past history. In fact, the effect depends on all possible histories of the particle. There are no everyday examples we can give to illustrate this phenomenon, but perhaps the following will suffice. Think of a wall being shaded by a screen with two holes in it. An electron beam that reaches the wall from a source on the other side of the screen is detected showing a wavelike interference pattern (see Fig. 7a). This interference pattern can only be explained by recognizing that the beam's history involves both holes. No individual electron can be said to go through one hole or the other; it can only be said that the beam goes through both holes.

Think of the universe as a particle (see Fig. 7b). Its present form is a pattern which can, perhaps, only be explained by taking into account all configurations that were possible in the past. These configurations

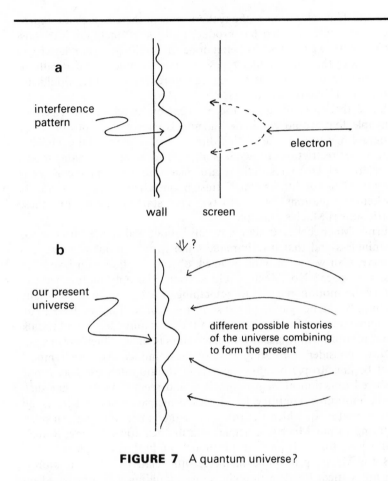

FIGURE 7 A quantum universe?

include the Bianchi types as well as all others, but you can't say the universe was ever "in" any one of these classes. What we are saying is this: The present universe is, in some sense, made of all possible universes, and that's the reason we have to study all possible universes.

Now, you might think that this is an impossible task, or that God is being malicious (or physicists stupid). We do admit that quantum cosmology is not the real reason people study these models. The main reason is that cosmologists are trying to understand Einstein's theory of relativity and its various ramifications. This endless theorizing might strike you, as it often strikes at least one of us, as being no more than a glass-bead game in which the real universe is of secondary concern. This relegation of the real universe (if there is such a thing) to secondary status is what has produced all these "metaphysical" cosmologies. However, work is being done to modify general relativity in such a way that these odder types of models would automatically be excluded. In quantum cosmology, therefore, these models would not contribute to the present universe.

Some theoreticians attack the problem by searching for a general principle that would eliminate all models except the isotropic ones. Progress has been made in this direction by Derek Raine, who has tried to set restrictions on what is allowed to be an acceptable model in relativity. These restrictions are meant to be a mathematically precise set of conditions which must be imposed on any solution of Einstein's equations of relativity. The starting point for these restrictions is Mach's principle.

Ernst Mach believed that any talk about "absolute space" was meaningless and that it only made sense to talk about what we can observe. Can we, as Newton claimed, observe motion with respect to absolute space? No, Mach would answer. The only motion we can observe is motion relative to something. That something is usually taken to be the body of the fixed stars, the nearest thing we have to an absolute reference frame. This notion that it is only meaningful to talk about motion with respect to the fixed stars is called Mach's principle.

Now, consider a rotating cosmological model. Any such model must be anisotropic because the axis of rotation defines at least one preferred direction at each point. It would seem, however, that such rotary motion is contrary to Mach's principle because there is no outside matter to which the rotation of the entire universe can refer. In trying to put Mach's principle on a mathematical footing, Raine's results were that only an isotropic model can fit the principle.

But is Mach's principle compatible with relativity? The question is still not settled. Einstein thought he was building a theory based on Mach's principle, a theory in which matter alone (e.g., the "fixed stars") determined the geometry. But it is clear that his theory allows many models—like the notorious Taub-NUT model we discussed—that have no matter in them whatsoever and yet still have geometry. The existence of this model could well be inconsistent with Mach's principle.

It seems, moreover, that if Mach's principle is correct, then the expansion of the universe itself should have some effect on local physics. As the reference frame of the fixed stars expands, local motion should become easier. After all, if it takes a certain amount of effort to move one percent of the distance between the two galaxies, it should require no more effort to move that one percent in the expanded-reference frame, even though the distance is greater. We realize this argument is nonrigorous, but there should be some effect of this kind. Perhaps the masses of things should decrease as the universe expands, making motion easier. Dennis Sciama, Robert Dicke and Carl Brans and P. A. M. Dirac have developed theories along these lines. For instance, the Sciama and Brans–Dicke theories tried to incorporate some aspect of Mach's principle by allowing the gravitational constant itself to be determined by the distribution of matter in the universe. Their main motivations were to provide an alternative theory against which general relativity could be tested. With this alternative theory we can decide which tests most sensitively distinguish between relativity and a viable alternative.

There have been recent tests (carried out by that strange kind of physicist known as the experimentalist) and relativity has won out over Brans–Dicke and all other contenders. At the end we seem to be left with our initial problem. By withstanding all the observational tests, relativity has burdened us with the task of explaining why the universe is so isotropic. We must also come to grips with Mach's principle and decide whether it is compatible with general relativity, a point not at all clear at the present time. Maybe there are other theories, like the Brans–Dicke theory, that provide a better framework for Machian cosmology. Or perhaps the real world does follow general relativity and even the far-out models we've talked about all contribute to a quantum theory of cosmology. One suspects, however, that Gödel's theorem, which states that a self-contained mathematical system must be either incomplete or inconsistent, may also apply to physics. If so, any theories we devise will all necessarily be unfinished.

A pessimistic vision? Perhaps, but at least it has the advantage that, if true, physicists will never be out of work.

2

Demythologizing the Black Hole

by Richard Matzner, Tsvi Piran
& Tony Rothman

If there is one phrase the public associates with the theory of relativity it is "black hole," an expression coined in 1968 by John Wheeler as he leafed through a thesaurus. In connecting the words "black" and "hole" to describe an object that had been studied peacefully by relativists for the better part of a century, he stumbled upon a profound psychological truth and unleashed a torrent of popular books and articles that has not ceased to the present day. Most of the expositions have been utterly nonsensical.

Alarmed at the attention received by the exotic idea of energy extraction from black holes, Tsvi Piran suggested we write an article to show how difficult the idea was to implement in practice. We did so and sent the result to Analog. *Meanwhile, Richard Matzner asked my advice on what to do with an article he had just finished that showed why black holes could not be used to travel from one universe to another. I suggested* Analog; *the editor there suggested that we combine the two pieces. This I did, adding some extra material as well. The article appeared in the September 1980 issue of* Analog. *We were pleased with the result, and apparently the readers were too: "Demythologizing" won* Analog's *annual readers' poll for best article of 1980 by a landslide.*

Tsvi holds a PhD from the Hebrew University Jerusalem (1976). His dissertation, "Astrophysical Processes near Black Holes," was in part concerned with energy extraction from black holes. Using sophisticated numerical models, he also investigates various types of gravitational collapse. He has done research at Oxford, Texas, and now divides his time between Jerusalem and the Institute for Advanced Study in Princeton. In addition to being a physicist, he is a radioamateur.

Richard Matzner received his PhD from the University of Maryland, where he worked with Charles Misner. He is now professor of physics at the Center for Relativity at the University of Texas in Austin. His research interests there include black holes and nonstandard cosmological models. He occasionally does observational work, and he participated in the 1973 eclipse expedition to Mauritania. He was also T. R.'s dissertation supervisor in Texas and, under the corrupting influence of his younger colleague, was last seen working on a science fiction novel.

"Demythologizing" first appeared in the September 1980 issue of Analog. *Having found only minor misprints, we reproduce it here essentially without alteration.*

Introduction: "Truth is for the Minority"

With the release of the Disney catastrophe, general interest in black holes has peaked. The release of *The Black Hole* also signals a critical overdose of misinformation to which the public has been exposed. We read statements like: " The pull of a black hole's gravity is so strong . . . time is stopped and space does not exist. . . . [A black hole's discovery] would unravel the mystery of both the universe's creation and eventual destruction."

Such blatant idiocy induces the public conception of black holes as monsters that gobble up all the matter in the universe, as miracle workers that can solve all our energy problems, as gateways to other universes and as time machines. This conception is profoundly misplaced. The same theory that predicts the black hole's existence also predicts that each of the preceding properties has severe limitations or does not occur at all. The very existence of black holes is itself debatable; within our own galaxy only one not-yet-conclusive candidate for a black hole has been found to this date—the X-ray source, Cygnus X-1.

Thus, it strikes us as bordering on the ridiculous to use black holes as an explanation for every property of known space. Of course, there are mistakes and there are mistakes. Some involve subtle points, and physicists advance their own field only by making lots of them. The layman cannot be faulted for doing the same. Nonetheless, most of the nonsense written about black holes stems from an ignorant exploitation of a sublime idea, and a lack of interest in the pursuit of knowledge. As we will see, the theoretical properties of black holes are in themselves so remarkable that there is no need to exaggerate them in an attempt to capture the public's attention. Bearing this in mind, we now examine some properties of black holes—without exaggeration.

Classical Black Holes and Canonical Misconceptions

Visualize a black hole. Most of us, encumbered by the limits of imagination, will visualize a small, black sphere floating in space among the stars. We probably think of this ball as a highly compressed solid, something like cold iron but unimaginably more dense. Unimaginably high density, we assure ourselves, produces an unimaginably great gravitational field. We further imagine the field to be

23

so strong that all surrounding matter is pulled into this tiny sphere, never to escape again. Light itself cannot avoid the same fate; fleeting, ephemeral, yet once light enters this strange object it is trapped forever by gravity. Thus, the "black hole"—absolutely black since light cannot be reflected from it to show its existence.

The question is, is this picture a description of anything? The answer is not straightforward but requires more precise concepts, caveats and "yes buts." In attempting an answer, one should first keep in mind that relativists, peddlers of gravitational theories, distinguish between several types of black holes. There is the basic Schwarzschild black hole, which is spherical, electrically uncharged and does not rotate; there is the Kerr black hole, which rotates and is not spherical; and there is the Reissner–Nordstrom black hole, which is spherical and nonrotating, but contains an electric charge. (The holes are named in honor of the mathematicians who worked out their theoretical existence.)

These three types, without additional complications, are lumped under the heading "classical black holes" to distinguish them from "quantum black holes." A quantum black hole is any black hole, including one of the above types, in which it is necessary to take into account the fact that light, for instance, consists of indivisible units called quanta. For light the quanta are photons; for the gravitational field itself the quanta are gravitons. Thus, we can have quantum Kerr black holes, quantum Reissner–Nordstrom black holes, and quantum Schwarzschild black holes. But for now we will limit ourselves to classical black holes.

Evidently, the above mental picture corresponds—more or less—to the basic Schwarzschild black hole. However, the emphasis in the previous sentence is on the "more or less," specifically on the "less." We will now begin to give a more accurate description of a classical black hole, keeping in mind that specific details may vary from one category of hole to the next.

The classical, astronomical picture of a black hole is one of a remnant left over by the collapse of a massive star; the examples typically used have about ten solar masses. The escape velocity from the surface of the black hole exceeds that of light; indeed, this is the definition of both a "black hole" and its "surface." The surface of a black hole is called the *event horizon*. Now, we know that no physical object can move faster than light, so nothing whatsoever, having fallen across the event horizon of a black hole, can come back out through that horizon.

A 10-solar-mass black hole has a radius of about 30 kilometers,

roughly the size of New York City. It is this typical example of a small, collapsed object with a gravitational field so strong that not even light can escape, that has conjured up the vision of black holes as extremely dense objects that grab anything in the vicinity. In fact, the density of the 10-solar-mass hole (density is the mass of the hole divided by the volume enclosed within the event horizon) is of order 10^{15} grams per cubic centimeter. This seems a very high density by everyday standards (the density of iron is only about 8 grams per cubic centimeter) until we realize it is comparable to the density in the nuclei of atoms. Each one of us is composed of particles of this sort of density.

In any case, a black hole does not have to be so dense. The basic black hole equations show a very simple relationship between the size and mass of a black hole and the density. As the radius of the hole or its mass is increased, the density goes down. Thus, by making a black hole large enough or massive enough, we can make the density as low as we want. Actually, there is no reason we could not make a black hole out of air. Such a hole would have a radius of about 30 billion kilometers, roughly 10 times the size of our solar system. When entering this black hole you would hardly feel a thing, but after a few days life would become uncomfortable—as you approached the singularity.

While we will not talk much about singularities in this article, we should mention that the singularity in the center of the black hole is the place where all the matter eventually ends up. The density at this point is infinite, which introduces a "yes but" into the above remarks. The density we have been discussing is the *average* density of the hole and, strictly speaking, one can only talk about the average density from outside the horizon.

At the surface of the air-bag black hole, the gravitational acceleration would be about 100 times the acceleration we feel on the surface of the earth, or roughly the same as the gravitational acceleration on the surface of the sun. A larger black hole, made out of hydrogen, would have an even lower surface acceleration. We see, then, that the gravitational acceleration of a black hole is not always overwhelmingly large.

If the gravitational field is so weak, the question immediately arises: why can't one escape by firing a rocket engine? The answer is somewhat tricky. We know that by accelerating even at very low accelerations—say, .1 gee—we can eventually reach huge velocities. Similarly, even the weak acceleration produced by our air bag will eventually accelerate objects to high velocity. In fact, by the time an object has fallen to the event horizon of any black hole, it is moving at the speed of light, inward. If the falling object wants to remain even stationary at the horizon, it must then move with the speed of light, outward. The principle of relativity says that nothing can move faster than the speed of light. Therefore, there is no escape. Acceleration is somewhat irrelevant to the problem; the speed of light simply cannot be exceeded.

Fancy Free: Orbits Around Black Holes

Related to the idea that a black hole possesses a strong gravitational field is the misconception that nothing can get remotely near the hole without being gobbled up. A good illustration of this nonsense is in the Disney film where the ship, the *Cygnus*, seems to require an antigravity field to prevent it from falling into the black hole. The filmmakers ignore the fact that at distances greater than about 10 times the radius of the black hole, ordinary orbital mechanics—known since the time of Newton—is applicable. For example, if the sun were suddenly replaced by a black hole of equal mass, the orbits of the planets would not change by the width of an ant's eyebrow. Admittedly, it would get dark, but that is another story.

This brings us to the first important rule of black hole orbital mechanics: *At large distances, the fact that we are in orbit around a black hole is irrelevant.* We may consider the black hole to be a spherical mass concentration producing an ordinary, Newtonian gravitational field, like that of the earth or the sun.

As the orbital radius approaches 10 black hole radii (300 kilometers for the 10-solar-mass case), general relativistic effects become very important. The proverbial *curvature* of space and slowing of time come into play. Such distortions of space and time manifest themselves in such effects as perihelion shifts in noncircular orbits. To understand perihelion shifts, we recall that Newtonian orbits are steady ellipses around the central body. The satellite's point of closest approach, the perihelion, remains at a fixed point in space. We say, in this instance, spacetime is flat or Newtonian. (See Fig. 1a.) When curvature of spacetime is more significant, the point of closest approach pivots around the central body with each orbit of the satellite. (See Fig. 1b.) This pivoting is called a "peri*helion* shift" when speaking of orbits around the sun, a "peri*astron* shift" when speaking of orbits around stars in general and a "peri*barythron* shift" when speaking of orbits around black holes. ("Barythron" is the Greek name for a deep pit in Athens into which condemned criminals were thrown.)

Because the shift is a cumulative, continuous effect, it can be

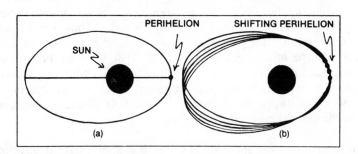

FIGURE 1 (a) A nonprecessing elliptical orbit. The perihelion, or closest approach to the sun, remains at the same point in space after each orbit. (b) A precessing orbit. The perihelion changes position slightly after each orbit around the sun.

detected even in satellites far from the central body, if a sufficiently long time is spent on the observation. For instance, Mercury's perihelion shift is about 42 seconds of arc per century—a very small effect indeed. We can say that, as far as Mercury is concerned, the spacetime curvature caused by the sun is hardly noticeable. Spacetime is very nearly flat. Close to a black hole, on the other hand, the peribarythron shift becomes very important. At 10 black-hole radii, it amounts to about 70 degrees per orbit!

Moving even closer to the black hole, circular orbits become unstable. A small deviation inward leads to a continuing spiral into the hole. For a Schwarzschild black hole, the point of instability comes at 3 black-hole radii (i.e., at 2 radii from the surface). This does *not* mean anything that falls within 3 radii of the black hole is irretrievably sucked in. One may still swoop down from a very large distance—down to 2 radii—and return to infinity, in the same way that a comet approaches and then recedes from the sun. And this approach can be made without engines, again, like a comet. If rockets *are* employed, one can come almost all the way down to the Schwarzschild radius, i.e., the horizon, and out again. Alternatively, one can continue to orbit around the hole below 3 radii; but, in this instance, rockets must be fired to maintain position. What is *not* allowed in this region are free, uncorrected orbits like those of satellites and skylabs around the earth.

In the Disney film, the featured hole was not Schwarzschild, but a rotating or Kerr hole (even though the computer graphics shown during the credits were mistakenly those for Schwarzschild). For a

Kerr hole, the point of the last stable orbit depends on how fast the hole is spinning, but the results are comparable to the Schwarzschild case; instabilities set in between 1 and 9 radii. Thus, the *Cygnus* should not need "antigravity devices" until very close to the hole indeed. On the other hand, as far as 1000 radii from the black hole, the *Cygnus* would be orbiting with a period of about one second. Admittedly, this may be why the antigravity device was posited in the first place—to dispense with orbits altogether. On the third hand, we doubt the filmmakers thought this far.

Since we have been speaking of orbits, it is appropriate at this time to introduce the second important rule of black hole orbital mechanics: *The principle of equivalence still applies.* This fact seems to have escaped the attention of almost all moviemakers and writers. The principle of equivalence states that any body in a free orbit or in free fall does not experience the force of gravity. We might say, "Falling free or orbiting 'round, equivalence says gravity not found." Examples of this are encountered in everyday life: When we dive off a diving board we feel weightless. When an airplane drops suddenly, those in it feel momentarily weightless. Astronauts in orbit around the earth are *not* weightless because gravity has been turned off above the atmosphere; rather, they are falling around the earth, continually diving off the board, if you will. Under these circumstances, the principle of equivalence says that gravity is not felt.

This is a very important point which applies to *any* situation near a black hole when rockets are *not* being fired: orbiting on a stable orbit, orbiting on an unstable orbit (when not correcting for instabilities), spiraling in, swooping down like a comet or just falling in. In these cases, one does *not* suddenly feel heavy near the hole. On the contrary, one feels weightless, as if he were orbiting the earth or diving into a swimming pool.

There is a complication to be introduced here. When an object comes close to a black hole, *tidal* forces can become extreme. As their name implies, tidal forces are those forces which raise tides on the surface of the earth. Because one side of the earth is slightly closer to the moon than the other side, the near side feels a slightly greater gravitational attraction to the moon than does the far side. Thus, we get a "tidal bulge"; the earth is stretched out in the direction of the moon. (Some readers may know there are actually *two* tidal bulges. We do not pause to discuss why this occurs.) We might say, with fair accuracy, that tidal forces are those which arise from the *difference* in the gravitational field between two points. The greater the difference, the greater the tidal forces.

Consider a man in a spacesuit orbiting a black hole. He is in free fall, so, by the previous discussion, feels perfectly weightless. However, the feet of the astronaut are slightly closer to the black hole than is his head. Therefore he experiences tidal forces: his feet are being pulled toward the hole more strongly than his head. As a result, the astronaut is stretched. One might think, because a man is so small, that the difference in the gravitational force between his head and his feet cannot be very large. After all, gravity does not decrease *so* fast over a couple of meters. This is not true. Near a white dwarf, neutron star or black hole, tidal forces can be immense. If he is orbiting a one-solar-mass body at a height of 10 kilometers, the tidal forces on our astronaut are approximately ten million times the force the earth is at this moment exerting on us. That is, while the earth is pulling us to its surface with a force which, by definition, is equal to our weight, the astronaut is being ripped apart by forces about 10 million times stronger. This particular example has roughly the conditions presented in Larry Niven's story, "Neutron Star." It is, alas, ludicrous to think the hero could save himself by curling up into a ball at the ship's center. More likely, he would end up spread over the walls, the consistency of pink applesauce. Perhaps, we have estimated, if he initially started out as a piano wire for triple high C, he might have survived.

Still, a caveat is in order here. Near black holes which are large enough (like our air bag), tidal forces become totally insignificant— much less than even those tidal forces we feel on earth. Thus, no shredding at all will take place near these holes until one falls close to the singularity. At the singularity, in all black holes, the tidal forces are infinite.

To sum up this section, we reiterate that it is the tidal forces which wreck spaceships near black holes, not the simple fact of strong gravity. And, as just mentioned, for very large holes (over about 10^5 solar masses), even this does not happen. As an astronaut orbits a black hole, he feels as weightless as if he were floating amid the clouds on a fine spring day. Near a typical black hole, though, his head is being wrenched from his feet by forces that make the bed of Procrustes amateurish by comparison.

Being and Nothingness:
Black Holes as the End of Space and Time

Two astronauts are orbiting a 10-solar-mass black hole. Richard, having seen one too many bad science fiction film, decides to end it all by taking the fateful plunge. He jumps. Tony, curious to see the

demise of his dissertation advisor, decides to clock Richard's fall to
the event horizon. "Time is on my side," Tony chuckles to himself,
but he has a surprise waiting for him. As Richard approaches the
event horizon, he seems to fall more and more slowly. Tony knows
this because Richard is carrying a green, flashing beacon. The time
interval Tony measures between each flash of the beacon is becoming
longer and longer. In addition, he is startled to find that the flashes
are growing much redder and dimmer "as time goes by." Tony grows
impatient; the fall seems to take forever. Tony dies of old age mutter-
ing, *"Veritatem dies aperit,"* but Richard has still not reached the
event horizon. Tsvi arrives in his space shuttle to take over the observations but suffers Tony's fate. He too grows old watching Richard's
beacon flash ever more slowly and redly. With his dying breath, he
entreats, "Stand still you ever-moving spheres of heaven/That time
may cease and midnight never come." Tsvi's descendents have no
better luck. Richard fades away completely just as he reaches the
horizon, after a truly infinite amount of time. The clock has stopped.

Richard, on the other hand, realizes, "Time and tide wait for no
man." He does not notice his beacon flashing any more slowly than
normal, nor does he notice it growing redder and dimmer. He reaches
the event horizon after a perfectly finite number of flashes. From that
point he crosses the event horizon, although he does not realize he has
done so, and continues his plunge to the singularity at the center of
the black hole. Of course, Richard is ripped apart by tidal forces long
before he gets there, but his dispersed atoms reach the dreaded singularity in a rather short amount of time—about 10^{-4} seconds as measured by his flashing beacon.

In addition to the problem of a mild discrepancy between two
clocks, there is a moral to this fable: Relativity is called relativity
because relativity is truly relative. The question, "Does time stop at a
black hole?," is meaningless as it stands. We can say, "To an observer
in a spaceship, an object falling into a black hole takes an infinite
amount of time to reach the event horizon." But we can also say,
without contradiction, "To an observer falling into a black hole, the
time required to reach the event horizon is quite finite." When posing
relativistic questions, we must be careful to specify about whom we
are talking or else we run the risk of lapsing into gibberish.

The slowing down of Richard's beacon-clock (as measured by Tony
and Tsvi on the ship) and the reddening of the light are two aspects of
the same effect. The curvature of spacetime associated with the gravitational field around the black hole actually causes time to flow at
different rates. Just as the flashing of the beacon can be thought of as
a clock, so can the oscillations in a light beam, or the movements of

atoms in the beacon motor. *Everything* is slowed down from the point of view of Tony or Tsvi on the ship. The slower oscillations of the light are interpreted by Tony's eye as a reddening of the light, and since light is being emitted from the beacon at longer intervals, fewer photons (light particles) reach the eye per unit time. The combination of these effects causes the excessive dimming of the beacon.

Richard, however, falling into the hole, is subject to the principle of equivalence. (Falling free or orbiting 'round, equivalence says gravity not found.) He does not feel any gravity on him or on his beacon. As far as he is concerned, there is no gravity to slow down his flashes and everything proceeds as normal, with the exception of tidal effects.

It is important to keep in mind that all these effects occur around any gravitating body—the sun, for instance. The only difference is in the magnitude of the effects, which will be much greater around a typical black hole than near the sun or the earth.

To conclude this brief discussion of space and time near a black hole, we would have wished to comment at length on the quotation found at the beginning of this article which says that a black hole is a place where "space does not exist." This, unfortunately, has proven to be impossible because we have failed to discover in that statement any meaning whatsoever.

The Cosmic Whirlpool: Kerr Holes and Penrose Processes

Present energy dilemmas have made popular the idea of extracting large amounts of energy from black holes. The attraction of this idea is not hard to see. We are all familiar with the large flywheels used by electric companies in their power plants. These huge flywheels store *rotational energy*. By coupling the flywheel to a generator, we are transforming the rotational energy into electricity for use in home and industry. In doing so, we have extracted the rotational energy from the flywheel and, as a consequence, it slows down.

Now, we have mentioned that Kerr black holes rotate, much like the above flywheels. The rotational energy of a rapidly rotating solar-mass Kerr hole is about 10^{54} ergs. At the earth's present rate of energy consumption, 10^{54} ergs would last approximately 10^{27} years, or about 10^{17} times the present age of the universe. This is a long time.

The question naturally arises, can the rotational energy of a Kerr hole be extracted? If it could, we would expect the black hole to slow down like the flywheel. When no further energy could be extracted, the black hole would no longer be a spinning Kerr hole; it would be a nonrotating Schwarzschild hole. In 1969, the British relativist Roger Penrose showed that extraction of the rotational energy of a Kerr hole is possible. Immediately after his suggestion appeared, others further proposed that the Penrose process might be used by an advanced civilization to tap the energy of black holes. From there, science fiction took over. The basic idea was used in *Gateway* by Fred Pohl. Indeed, one of us (T. R.) succumbed to the temptation to use the idea in his novel, *The World Is Round*. Unbeknownst to T. R., T. P. and others were at the same time proving how difficult the Penrose process was to implement.

To understand the Penrose process further, we must first talk in more detail about Kerr holes. The rotation of a Kerr hole causes a "whirlpool in space." This whirlpool is actually quite similar to an ordinary ocean whirlpool except that, instead of water whirling around, it is spacetime itself swirling around the black hole. If a space traveler is caught in this whirlpool, he is dragged around the black hole exactly as he would be dragged around the eye of the vortex if caught in an ocean whirlpool. If the space traveler wanted to remain stationary, he would have to fire his rocket engines to overcome the spacetime dragging. Again, this has a marine analogy. A swimmer must swim against the current in the vortex if he wishes to remain in the same place.

We should note that this dragging is not unique to black holes but, according to relativity, occurs around any rotating body. In fact, a team of experimentalists at Stanford, led by Francis Everitt, is planning to measure the dragging force caused by the *earth's* rotation. This measurement will be carried out by a satellite to be launched by the space shuttle. The dragging caused by a tiny body like the earth is really very small. While the Stanford satellite orbits the earth, the gyroscopes on board will be tilted a slight amount by the drag. After a full year, the cumulative angle of tilt will be less than a second of arc—about the angle subtended by a penny as seen from a distance of a kilometer.

Although the effect due to the earth is small, around a black hole the dragging can become enormous. In fact, beneath a certain distance from the hole that is termed the "stationary limit," no matter how hard one fires his rocket engines against the current, the dragging cannot be overcome and one is inevitably swept around the hole. This notion can be made more precise. Consider an observer on a "space buoy" being dragged passively around the hole. To him, someone in a rocket trying to overcome the dragging will appear to be moving in the opposite direction. At the stationary limit, this rocket will appear to the observer on the buoy to be moving at the speed of light. From a space station far above, however, the rocket is just managing to fight the current and remain stationary, hence the name "stationary limit."

We recall the famous words of the Red Queen: ". . . it takes all the running you can do to keep in the same place. If you want to get anywhere else, you must run at least twice as fast as that." Unfortunately, one cannot run any faster than the speed of light. If she is unlucky enough to fall beneath the stationary limit, even the Red Queen will never be able to stay put and will be dragged around the hole along with space buoys, rockets and everyone else. (See Fig. 2.)

The region between the stationary limit and the event horizon is called the "ergosphere." "Ergosphere" was coined by Wheeler and Ruffini from the Greek word "ergon" meaning "work." It is in the ergosphere that the Penrose process takes place. (See Fig. 3 for the relationship between the horizon, ergosphere and stationary limit.)

FIGURE 2 An observer on a space buoy, being passively dragged around the black hole whirlpool, sees the Red Queen at the stationary limit running at the speed of light. From a space station far above, however, she is seen as just managing to remain at the same place. (Note: This drawing is not to be taken too literally. One does not actually see a spacetime whirlpool around a black hole.)

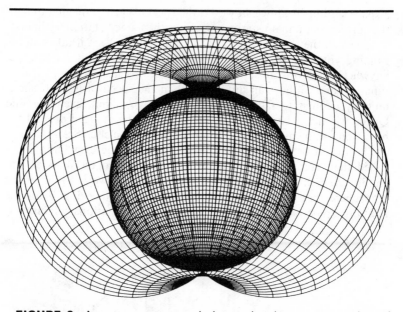

FIGURE 3 A computer-generated picture showing a cutaway view of a Kerr hole rotating at the speed of light. The outer surface is the stationary limit; the inner surface is the event horizon itself; and the region in between is the ergosphere. The view is from 26.5° above the equator, which accounts for the slight distortion of the inner sphere. We wish to thank Nigel Sharp for generating this picture. (He points out that all previous textbook views of Kerr holes in Boyar-Lindquist coordinates without the cusps shown here are incorrect.)

Consider a rocket orbiting in the ergosphere. It ejects a load of garbage *against* the current (like the Red Queen's movement). Although this garbage is swept around the hole—since it *is* beneath the stationary limit—it is "struggling against the current." One can imagine that an object moving on such a "counterrotating" orbit would exert a braking force on the hole and slow it down in the same manner as we slow a flywheel. Thus, some of the rotational energy is lost and is, in fact, transferred to the ship as a recoil effect. (Think of a gun shooting a bullet. The recoil is greater than normal owing to the presence of the rotating black hole.) This energy would be manifested as a greater kinetic energy of the ship—that is, a higher velocity. The ship could then leave the ergosphere with more energy than it had to start with, to be used elsewhere.

However, the matter is not so simple. The ejected garbage will be captured by the black hole, adding its own mass to the original mass of the hole. Since $E = mc^2$, by losing the garbage we are losing energy to the hole. If the amount of energy lost to the hole is greater than the amount of energy gained by braking the hole, we have a net loss of energy. No extraction has taken place.

Nonetheless, if certain conditions are met, the energy balance will be favorable. That is, if the garbage is ejected at sufficiently high velocity onto a counterrotating orbit within the ergosphere, the net result will be an energy gain. Any orbit which meets these requirements is termed an "energy extraction orbit." We emphasize that they only exist within the ergosphere. Note also that it is the orbit which is important, not what we eject. Therefore, it makes sense to use garbage, since this eliminates waste disposal problems as well.

The Penrose process is best illustrated by the machine in Fig. 4 (adapted from the text *Gravitation*, by Misner, Thorne and Wheeler).

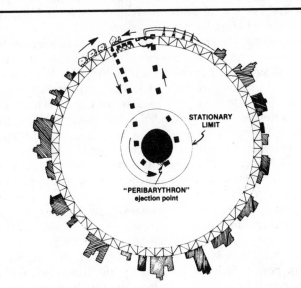

FIGURE 4 The black hole energy extraction machine. Shuttles from the shell enter the ergosphere, jettison garbage onto an energy extraction orbit, return to the surface with a gain in energy to power a giant generator. The additional energy comes from the energy content of the garbage and from the rotational energy of the black hole. In this process, the hole slows down.

From GRAVITATION by Misner, Thorne and Wheeler, W. H. Freeman and Company © 1973.

An advanced civilization builds a huge shell around a Kerr black hole. Space shuttles loaded with garbage enter the ergosphere. They eject their payloads in the manner already described, receive a giant dose of energy which boosts them to huge velocities and then return to the surface. They are caught in the arms of a giant generator that converts this kinetic energy into electricity for use over the shell.

Even if our supply of garbage is limited to 1 earth mass, this is enough to power the Penrose process for about 10^{21} years, or 10^{11} ages of the universe—not an insignificant amount of time.

Two technical details make this extraordinary picture somewhat less optimistic. The first difficulty is jettisoning the garbage onto an energy extraction orbit. The second difficulty is getting the boosted shuttle out of the ergosphere without being captured by the black hole. We can better understand these problems if we pretend we are on a shuttle, the *Penrosia*, whose mission is to go into the ergosphere, dump garbage and return to the shell with as much energy as possible. The crew is fresh out of Starfleet Academy and so learns by the dangerous method of trial and error.

We have entered the ergosphere. Because fuel supplies are limited, the Captain has turned our engines off. The *Penrosia* is now being passively dragged around the black hole's whirlpool like a space buoy. Since we are in orbit, we feel weightless. An inexperienced space cadet attempts to eject a load of garbage onto an energy extraction orbit simply by throwing it out by hand. To our dismay, we find that the garbage only follows the shuttle along, very gradually drifting away (exactly like what happens with garbage jettisoned from a space capsule in earth orbit). The bundle is certainly moving too slowly to be on an energy extraction orbit; this garbage would hardly brake a snail, let alone a black hole. Determined, the crew tries again, this time firing the garbage out of a cannon. Now the garbage vanishes into the distance, but when our sensors plot the trajectory, we find that the garbage is still not moving fast enough to be on an energy extraction orbit. Many such attempts are made, each using increasing amounts of power. They all fail. Finally, the frustrated crew of the *Penrosia* succeeds in shooting a thimbleful of garbage onto an energy extraction orbit. They calculate the velocity of the thimble and find it to be nearly the speed of light. This has been accomplished only by momentarily diverting the full power of the shuttle's reactor engine, just for the purpose of launching the thimble. When the energy balance is computed, the crew discovers that the energy generated by the reactor on board was almost equal to that gained by ejecting the garbage. They have gained some energy, however, and tired but happy, prepare to leave the ergosphere.

At this moment, the Captain realizes he has made a fatal mistake: he has forgotten Newton's third law. When a rocket ejects fuel from its engines, the rocket is propelled in the opposite direction. By ejecting garbage from the *Penrosia*, the crew has inadvertently boosted the shuttle onto a new orbit. The ship's computer makes a quick calculation. To the Captain's horror, he realizes that the new orbit will lead the *Penrosia*—and us—directly into the black hole. The Captain guns his engines. After expending all the energy gained by launching the thimble, we barely escape to the surface, tired but unscathed.

What happened?

Recall our previous discussion. The "sufficiently high velocity" mentioned earlier for an energy extraction orbit, turns out to be nearly the speed of light. That is, the garbage must be like the Red Queen, moving at nearly the speed of light with respect to other objects in the whirlpool. To accelerate an object to the speed of light requires stupendous amounts of energy, which, in this case, must be generated on board. It turns out that we must convert a large part of the garbage into energy in order to boost what little remains to the velocity of light.

The second problem was to get the energy out. Most orbits within the ergosphere intersect the black hole. The *Penrosia* boosted herself onto one of these orbits and to escape it required a stupendous amount of energy. In most cases, anything gained by the ejection is lost in trying to escape. This second problem, as it turns out, can be overcome only by jettisoning the garbage exactly at the peribarythron of the orbit. We then escape to the surface with what little energy was initially gained by the ejection.

We have just seen that to get an appreciable amount of energy out of the hole requires that matter be converted on board the shuttle with essentially 100% efficiency. Then by ejecting this "energy beam" (photons), we get an additional 20% boost from the hole, for a grand total of 120% efficiency. Not bad; but since this requires almost 100% conversion efficiency to begin with, the Penrose process might not be worth all the trouble. It can, however, be used for a more efficient energy conversion process than is available on earth. That is, it turns out we can use a modified Penrose process to extract energy from matter with up to 10 times the efficiency of the 0.5% of hydrogen bombs, the most efficient process known at present. Unfortunately, we do not have space in this article to discuss such modifications, and a more detailed discussion will have to wait for another opportunity.

"I Expect to Pass through This World but Once": Star Gates and Time Machines

Any interstellar empire or commercial consortium needs a means of rapid communication and transport. In *Star Wars*, the smuggler Han Solo made the "jump into hyperspace" and emerged at his destination some time (12 parsecs!?) later. Space warps and star gates are a staple of science fiction.

Relativity, as already mentioned, describes gravity as a warping of space and time, and a black hole is the result of the strongest possible curvature. It is not surprising, then, that science fiction has latched onto black holes in an attempt to make space warps sound more plausible. To some extent, it is the fault of relativists; in idealized situations, we have discovered the tantalizing possibility of a "star gate" lurking in black holes. Unfortunately, the situation has gotten out of hand and almost everyone has chosen to ignore work started as far back as a decade ago by Penrose and Floyd that shows that "star gates" cannot be realized in practice.

To discuss this problem, we will need to back up and fetch some concepts not yet introduced in this article. Relativity is, in a sense, a study of geometry, but not simply the ordinary Euclidean kind that we all learn in high school. For one thing, space and time have been combined into a four-dimensional spacetime. To pursue this point briefly, let us refer to Fig. 5. Here, only two spatial directions are shown, x and y (east–west and north–south if you like) and the time direction, labeled by ct. Time increases upward. From the explanations accompanying Fig. 5, we distill four rules for understanding these diagrams. (1) An object stationary in space still moves through time. Its path through spacetime, or *worldline*, is therefore a vertical line. (2) An ordinary, moving object, like a rocket, has a worldline that is tilted at less than 45 degrees from the vertical. (3) Light travels along 45-degree lines. (4) Traveling on a worldline tilted greater than 45 degrees from the vertical is prohibited because this is motion faster than the speed of light.

This type of spacetime diagram has its defects. The most serious one is the difficulty of showing things that are very far apart—it is especially difficult to map an infinite universe onto a finite piece of paper. Nonetheless, with sufficiently vigorous squeezing, we can actually distort the outer edges of the universe in such a way that we can fit the entire infinite universe onto a finite piece of paper. We can even retain certain features of the real universe. The one feature that is usually kept is the 45-degree angle that represents the trajectory of a light beam. A diagram like Fig. 6 results.

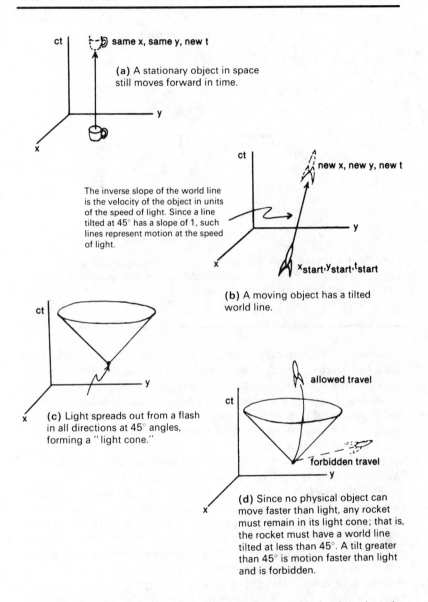

ct

`same x, same y, new t`

(a) A stationary object in space still moves forward in time.

y

x

The inverse slope of the world line is the velocity of the object in units of the speed of light. Since a line tilted at 45° has a slope of 1, such lines represent motion at the speed of light.

ct

`new x, new y, new t`

y

x

$x_{start}, y_{start}, t_{start}$

(b) A moving object has a tilted world line.

ct

y

x

(c) Light spreads out from a flash in all directions at 45° angles, forming a "light cone."

ct

`allowed travel`

y

`forbidden travel`

x

(d) Since no physical object can move faster than light, any rocket must remain in its light cone; that is, the rocket must have a world line tilted at less than 45°. A tilt greater than 45° is motion faster than light and is forbidden.

FIGURE 5 Basic spacetime diagrams. The time scale is plotted vertically. The label "*ct*" is just to get the units correct.

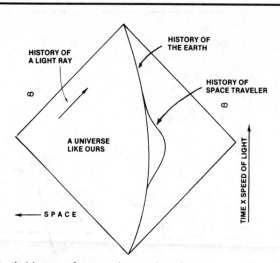

FIGURE 6 A history of our universe (much compressed around the edges). The space traveler's history looks longer, but according to the curious geometry of spacetime, actually amounts to a shorter time than passes on Earth between his departure and return.

All this is in preparation for a discussion of star gates and time machines based on black holes. Recall our discussion of tidal forces. We mentioned that in a simple Schwarzschild black hole, tidal forces on an infalling object (remember Richard's plunge) become greater and greater until they become infinite at the singularity. Well, inside a rotating Kerr hole, or a charged Reissner–Nordstrom black hole, this is not necessarily true. In fact, the theory states the following: *Unless the black hole has exactly zero charge and zero rotation, it will allow an object—say a spaceship—whose own gravitational effects are negligible, to enter the black hole at a speed slower than light, avoid the singularity, and leave again by an exit different from the surface through which it entered.* The different exit surface gets around the immediate objection that nothing that falls into a black hole can escape again; it can, but not through the same surface through which it entered.

If the exit surface is not the same as the event horizon by which we entered, then where does the spaceship end up? Fig. 7, a spacetime diagram drawn for a Reissner–Nordstrom black hole, attempts to explain the situation. The two diamond-shaped regions are like Fig. 6; thus both represent infinite universes. The diagram therefore shows two universes connected by a black hole tunnel. (Again, we are plotting space horizontally and time vertically.) The curve shows the history of the infalling-spaceship-cum-observer as it travels through the hole.

A complicated figure indeed. The intrepid traveler falls inward. The first 45-degree line he crosses is the event horizon of the black hole; nothing can re-exit via this surface once having crossed it. The jagged lines represent the singularities, where tidal forces are infinite. (Remember, singularities are moving forward through time; therefore on this diagram they appear as vertical lines.) But the traveler can avoid these reefs; just by coasting he passes at a safe distance. When he emerges from the black hole, he finds himself in a normal universe like the one he just left. In fact, it may be precisely the one he left, but the black hole exit need not be near the entrance, and there is reason to think it would not be near that entrance.

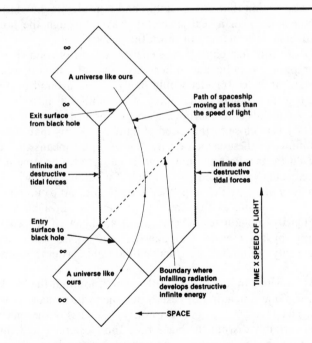

FIGURE 7 A spacecraft might fall into a black hole and hope to emerge via the star gate into a universe like ours—maybe in fact ours. (See Fig. 3.) The dotted line is where infalling radiation can become destructively large and close this star gate, according to the general theory of relativity.

We might at first worry whether the second universe is the same as ours. According to the theory, there is no reason it should not be, but equally no reason it should be, either. (Parallel worlds!) If many charged or rotating black holes inhabited our universe, the possibility would exist that their exits would emerge in our own universe, or all in the same second universe, or in any number of alternate worlds (see Fig. 8). More bizarrely, these various universes might be connected in such a way that, after traveling through several black holes, we return to our own universe, but at a previous time. The theory does allow for this possibility, suggesting the use of black holes as time machines. In any case, the first pioneer to determine which of these possibilities exists will be a very brave man, and exceedingly dedicated to the progress of science.

Unfortunately, at this point, hard science drags us back from such interesting speculation. The crucial flaw in the above discussion was the assumption that the spaceship had a negligible gravitational effect on the hole. Can a real spaceship travel through the hole without disturbing its structure? In short, the answer is no.

To see this, just consider the energy content of solid space-garbage, laser photons, radio and other waves, all of which a well-equipped interstellar space traveler would likely spread around during his trip. The central problem is that some of this stuff falls into the black hole, too. As it falls into the hole, garbage, for instance, has picked up a velocity very close to the speed of light. We know that mass increases to infinity at these velocities. By $E = mc^2$, this means that the energy of even the smallest amount of garbage has also become infinite. The same occurs with the energy content of infalling radio waves and light signals emitted by the ship; their energy goes up to infinity, also. But what has infinite matter and energy densities in a black hole? The singularity, of course. Where the clean black hole provided clear sailing, by allowing for garbage or radio waves we have created a singularity of infinitely destructive tidal forces, as in the Schwarzschild case.

A moment's thought leads to the conclusion that the black hole star gate is a one-time affair at best. If just the radio signals transmitted by the space traveler can be so disruptive, the mass of the spaceship itself must certainly disrupt the black hole and close the gate behind him— assuming he makes it through in the first place. What if he is extremely careful not to sully the black hole before his journey? He maintains radio silence and stows his garbage bags. Could he make it through the tunnel?

Surely, his spaceship must have substantial mass. A mass falling toward a black hole is accelerating and, consequently, generates *gravi-*

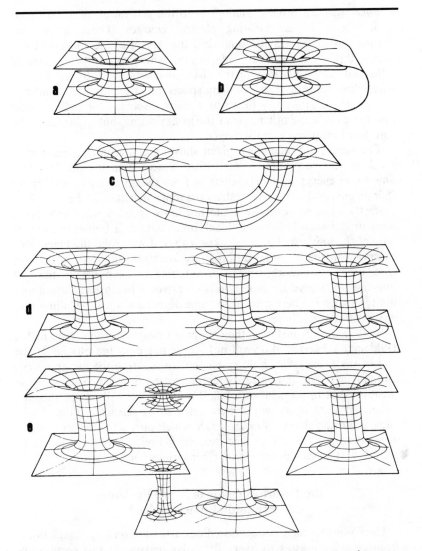

FIGURE 8 (a) A schematic picture of a black hole connecting two disjoint universes; (b), (c) The two universes may in fact be the same; then the distance between the entrance and exit *might* be much shorter by the hole. (d) If the two universes in (a) are different, we would hope that at least every connection would be between the same two universes. Appealing to Murphy's law, (e) is much more likely: a completely tangled web of interconnections.

tational waves, a process analogous to the generation of electromagnetic waves by accelerating electric charges. These waves are oscillations in the gravitational field that travel at the speed of light. Again, this is analogous to radio waves, which travel at the speed of light and are oscillations in the electromagnetic field. Some of these waves travel ahead of the infalling spacecraft, are amplified to infinite energy in the same way that was already discussed for radio waves; and the gate to the other side of the galaxy slams shut in his face. The star gate cannot even be used once.

The simple argument just given shows that the inner structure of black holes is unstable to small disturbances from the outside. Any amount of energy or debris falling in from the outside will develop an infinite energy density and destroy the inner structure of the hole. In a perfectly clean universe, we could not know *a priori* whether a given Kerr or Reissner–Nordstrom black hole contains a tunnel to another part of the universe. But we do know that, if we probe the black hole by trying to reflect radiation off it, we automatically destroy the star gate, because some absorption of radiation is inevitable. In the real universe, of course, the situation is even worse because radiation and matter are everywhere present to some degree and must be falling into any existing black holes.

So it seems, for instance, the collapsar transport system used in Joe Haldeman's *Forever War* will not work; nor does the physics at the end of the Disney film hold water (not to mention the metaphysics); and, in fact, the siphoning of extra matter from another universe mentioned at the conclusion of T. R.'s own novel will not go through either. At least these works were released as science fiction. Books such as Adrian Berry's *The Iron Sun*, which purport to be science, are either flagrant rip-offs or bad science fiction. As either rip-offs or science fiction, they deserve no further serious consideration.

Big Surprises Come in Small Packages: Primordial Black Holes

Until now, we have concentrated our attention on large black holes, from 10 solar masses to over 10^{10} solar masses. In this section, we turn our attention to the other end of the spectrum: mini black holes. At the very beginning of the universe, at times much less than one second after the big bang, the density of matter was comparable to what is found in typical black holes. It is conceivable, then—but by no means proven—that a slight fluctuation in density would "snap" the matter into black holes. Such "primordial" black holes would range in size from the very large, about 100,000 solar masses, to the very small, about 10^{-5} grams. The small holes would be formed first, when the density was highest, followed by successively larger holes, until the density was too low to form any at all.

Large primordial black holes would behave in exactly the same way as the other large holes that we have already discussed. There is nothing to be added here. At the other extreme, holes of 10^{15} grams and below are remarkable objects. The density of 10^{15}-gram black holes is so high that 1 cubic centimeter of them would contain the known mass of the universe! Holes smaller than this mass would exhibit extraordinary quantum properties, specifically the famous Hawking radiation named after its discoverer. Space does not permit us to discuss these amazing properties. Suffice it to say, there is no observational evidence to indicate that holes smaller than 10^{15} grams exist or existed. Moreover, theoretical upper limits placed on such holes by Page, Hawking, Novikov *et al.*, and two of us (R. M. and T. R.), indicate that, if they ever existed, they were few and far between. For instance, there cannot now be more than about ten black holes of 10^{15} grams per cubic parsec, each with the mass of a mountain but the size of a proton.

Primordial black holes with masses greater than 10^{15} grams have negligible quantum properties and can be treated classically. Such black holes have also received attention in science fiction and popular folklore and, therefore, their share of misrepresentation. Perhaps the most famous—or notorious—suggestion was put forth by Al Jackson and Mike Ryan, then at the University of Texas, that the 1908 Tunguska blast in Siberia was caused by the collision of a 10^{21} gram black hole with the earth.

We may first ask, "What are the odds of such a collision taking place?" Not bloody likely. Assuming all the observable mass in the galaxy to be concentrated into 10^{21}-gram black holes, we can calculate that one collision should occur about every 10 ages of the universe, or 10^{11} years, Marauding black holes do not seem an overwhelming threat to U.S. security. Nonetheless, it is possible that Tunguska was *the* collision. Although a 10^{21}-gram black hole is small in radius, about 10^{-7} centimeters, its mass is large, about one million small mountains. Jackson and Ryan proposed that the gravitational attraction of this hole caused the surrounding air to be yanked inward, resulting in a compact ball of air whose shock effects produced the destruction seen in well-known photographs. There was, however, substantial debate on whether a black hole of this mass would have the claimed effect when interacting with the solid earth. Most physicists believe the ground shock would have been tremendous, something like an H-bomb detonation for every meter of travel.

So in scientific circles the matter is considered dead and buried. In any case, Al Jackson and Mike Ryan have on occasion confided that the suggestion was not entirely serious in the first place.*

Detailed statements about the interaction of smaller black holes (about 10^{15} grams) with matter are difficult to make. Nonetheless, simple calculations give the following general picture, which should not be too far wrong. Recall, a 10^{15}-gram black hole has the diameter of a proton. This is too small a size to rapidly accrete (gobble up) surrounding matter. Even if we consider that any nearby particle in random motion falls in when nearing the hole, we find an accretion rate such that the black hole will not even double its mass in the lifetime of the universe. Talk of eating a planet becomes absurd. Thus, the black hole posited by Larry Niven in "The Hole Man" would certainly never gobble up Mars in less than extreme cosmological times, meaning millions or billions of ages of the universe.

Where Do We Go from Here?: Prospects for the Future

We have talked about many types of black holes and many of their properties but have omitted discussion of many other interesting properties. We have not spoken about Hawking radiation, nor about black hole collisions, nor about superradiance, nor about astrophysical accretion, nor about photon trajectories and imaging properties, nor about the influence of primordial black holes on nucleosynthesis after the big bang. We have also shied away from direct discussion of the famous singularity that occurs within all black holes. The singularity, as already mentioned, is the center of the black hole into which all the matter has fallen. It is a place where the density of matter, as well as gravitational and tidal forces, are infinite. When people speak of space and time ending at black holes, they are perhaps thinking of the singularity. But it may be a mistake to say space and time end at the singularity; what ends is our present knowledge of physics.

Much work is currently underway to remedy the situation. Many physicists believe that true singularities do not exist, that at such small distances the quantum properties of spacetime itself come into play. They suggest that matter cannot be compressed to a smaller size than the so-called Planck length, about 10^{-33} centimeters, where the quantum effects become dominant. According to this view, the singu-

---

* We have recently learned that a report in *Sotsialisticheskaya Industriya*, 1/24/80, indicates that ordinary meteoric debris was recently found at the Tunguska site.

larities of classical black holes are nonexistent in reality and are only the temporary nuisances of defective mathematics.

At least two Russian physicists, Frolov and Vilkovisky, have recently claimed to have proven that black holes, in some sense, do not exist at all. Proper use of quantum field theory, they argue, shows that as matter collapses to form a black hole, it misses the singularity, "rebounds," and eventually re-expands beyond the event horizon. This process, for even Tunguska-sized black holes, will take longer than the age of the universe. Nonetheless, in a strictly logical sense, a black hole is no longer a black hole, but only temporarily out of sight. We are not yet sure whether Frolov and Vilkovisky are correct, but we are certain that the full merger of relativity and quantum theory will reveal many answers and even more questions.

3

Grand Illusions:
Further Conversations
on the Edge of Spacetime

by Richard Matzner, Tony Rothman
& Bill Unruh

By the time "Demythologizing" went to press, a number of new articles had appeared in well-known magazines. These articles, to put it mildly, did violence to our sense of physical law and fair play. It is one thing to make honest mistakes; it is quite another to engage in deliberate deception. One should not write on subjects about which one knows nothing. In the group session that witnessed the savaging of the above-mentioned articles, one of the most adamant participants was Bill Unruh, a professor of physics at the University of British Columbia in Vancouver. He received his PhD from Princeton under John Wheeler, and his research is concentrated in the field of quantum gravity. Among his many contributions, the best known is the discovery of the "Unruh vacuum," which is of fundamental importance in understanding such processes as black hole radiation.

The article that resulted from our discussions took over a year to get written. (An appendix, "Communication with Observers Falling into Black Holes," appears at the end of this article. Its co-author, A. C. Ottewill, is a fellow at Merton College, Oxford University.) We had hoped that "Grand Illusions" would appear in Analog *as a sequel to "Demythologizing," but owing to circumstances beyond our control, this did not take place. Here, then, is its first printing.*

●

I. Introduction: The Struggle Continues

Since our last excursion into spacetime, black hole apocrypha have continued to multiply at an alarming rate. Unfortunately, no new film featuring a black hole has been released to spur us on to ever greater heights of spleen and, consequently, of eloquence. There have surfaced, however, a number of articles—purporting to have some connection with science—that need to be dismembered, and a number of persistent misconceptions that should be interred once and for all. As in "Demythologizing the Black Hole," we will wind our way through some involved arguments, but because you coped so well in the first round, we are confident you can survive the second.

Our audience will undoubtedly notice that many of the statements made here stand in opposition to those found elsewhere in the popular literature. The title of this article and that of its predecessor indicate that we are well aware of the conflict. In science, a worker expects criticism for a faulty job. For this reason we do not hesitate to point out errors where we find them. Most of our pronouncements are based on calculations made by colleagues or ourselves, and we are confident of them. Nonetheless, it is easy to fall into delusion when working on the edge of spacetime, and we expect to be told if we've gotten something wrong.

Before dealing with the new harvest of gravitational disturbances, we would like to take up some old business. The reader response to "Demythologizing" after it appeared in *Analog* was indeed gratifying. Some points of general interest were raised that deserve comment. For the purpose of this discussion as well as the purpose of comprehending the remainder of the article, it would be helpful to review some standard terminology.

II. Clearing Cobwebs: Review

A standard 10-solar-mass black hole may be thought of as the remnant of a star that has used up all its nuclear fuel, has thereby lost the internal pressure needed to support itself and has consequently collapsed under its own weight. The collapse has compressed the star enough so that the escape velocity from its surface exceeds the velocity of light. Because the velocity of light cannot be exceeded, no object can escape once it has fallen within the black hole, and it must plunge to the center of the hole, a point called the *singularity*. At the singularity, all gravitational and tidal forces become infinite. The "surface" of the black hole itself is termed the *event horizon*, and another way of saying that a black hole has formed is to say the star has collapsed to within its event horizon. One should not, however, think of the event horizon as a solid surface—it isn't. It has more the character of the edge of a waterfall. There is no wall at the juncture, but once passed, return is impossible.

The event horizon is located at a characteristic distance from the singularity. For a nonrotating black hole, this distance is known as the Schwarzschild radius and is given by the formula $R = 2GM/c^2$, where G is the gravitational constant, M is the mass of the hole, and c is the speed of light. A nonrotating hole is called, not surprisingly, a Schwarzschild black hole in honor of Karl Schwarzschild, who worked out the theoretical formalism in 1916. The shape of a Schwarzschild black hole is exactly spherical and has the radius given by the Schwarzschild radius. To give a feel for the size of the numbers involved, a 10-solar-mass black hole has a radius of about 30 kilometers.

If the black hole is rotating, the configuration is no longer spherical, but has a size roughly equal to that of the Schwarzschild black hole. The rotating hole is usually referred to as the Kerr hole, in honor of Roy Kerr, who worked out the mathematical description in 1963. Because most, if not all, stars are rotating, most black holes of astrophysical interest are expected to be Kerr holes.

We will sometimes use the words "classical" and "quantum" when referring to black holes. Unless otherwise stated, you can assume that the type of black hole being discussed is a classical one, a hole that does not require quantum mechanics for an accurate description. Discussion of quantum holes will be deferred to the last section. Finally, since the advent of $E = mc^2$ seventy-odd years ago, physicists have grown accustomed to using the terms "mass" and "energy" interchangeably. In only a few instances in this article will the distinction be important, and we will try to keep this clear. Otherwise, you must forgive us for lapsing into the relativist's habit of mentally setting the speed of light equal to 1, so that energy and mass are exactly equivalent.

III. Old Habits Die Hard:
Comments on Our Readers' Comments

The episode in "Demythologizing" that caused by far the most uneasiness concerned Richard's plunge from a spaceship into a black hole. Many readers simply refused to believe that to Richard—falling into the hole—only a finite amount of time separates him from oblivion, while for Tony—orbiting in a ship above—Richard takes an infinite amount of time to reach the event horizon. (You should not feel too bad if you found this subject perplexing, for you are in good company; our own Dr. Asimov also got it backwards, but we have since set him straight. Such is the learning process.) Most of the objections to our discussion were rather vague, but seemed to assume the existence of some "absolute time" that would necessarily be the same for both Richard and Tony. The absolutist approach was best exemplified by one reader whose objection ran thusly: "I agree that Tony sees Richard's clock stop, but to say that it takes an infinite amount of time for Richard to reach the event horizon is nonsensical."

This type of intuition has plagued relativity since its inception and is responsible for all the confused discussion concerning, for instance, the famous twin paradox. The point is that you can *only* measure time with a clock, and "the time" is exactly what your clock says—assuming it is not broken. To talk about some "absolute time" that does not refer to a clock is meaningless. Could you talk about days and nights without the sun? If Tony sees Richard's clock slow down and finally stop altogether, he can only say that an infinite amount of time is elapsing between ticks. Similarly, if Richard sees his clock behave normally, he can only conclude that time is advancing at its usual rate.

Proceeding from this discussion, one astute reader thought he had caught us in a contradiction. If, from Tony's point of view, any matter falling into the black hole takes an infinite amount of time to reach the event horizon, then how can Tony say a black hole exists in the first place? That is, assume the star begins to collapse as Tony watches. To form a black hole, the surface of the star must collapse to within the event horizon. But because the star itself is just infalling matter, surely this "gravitational collapse" must also take an infinite amount of time as seen from Tony's spaceship. So how can he logically claim that a black hole forms at all?

Like many answers in physics, there are several complementary and equally correct responses. First, let us again assume that the star has begun to collapse just before Richard jumps, and while Tony watches. From Richard's point of view, the matter of the star is falling just ahead of him, at the same rate that he is falling. The time he measures for the star to cross the event horizon is exactly the same amount of time that it takes him to cover the same distance. In the frame of Richard and the collapsing star, the event horizon is crossed after a very short amount of time. Hence, there is no escape to the ship above and a black hole has truly formed.

From Tony's point of view, on the other hand, the star collapses to within a few millimeters of the event horizon on a time scale typically of a few seconds and, for the purpose of most discussions in astrophysics, whether it takes an infinite amount of time to travel the remaining millimeters is irrelevant. Nonetheless, there is a real sense that the star has formed a black hole, even as seen from the spacecraft. Recall, in our last exciting episode, Tony observed Richard's progress by watching a flashing beacon that Richard carried in his fall. Suppose now that the beacon fails. Tony can still see Richard by reflecting the ship's searchlight beam off his erstwhile companion. But Tony is so distraught that he does not at first think of this possibility. It is only one orbit after Richard has jumped—following an on-board screening of *The Black Hole*—that Tony thinks to switch on the beam. The photons from the searchlight chase Richard toward the hole but his headstart is so great that, from the ship's vantage point, they never reach him; Richard has disappeared forever, as has all the matter collapsing with him into the hole. A more precise way of viewing this is to consider the events from Richard's reference frame. The photons catch up with him only after they and he are within the event horizon, so they cannot escape again for Tony to see them.

Thus, there is a critical time associated with the collapse of a star or a falling astronaut after which there is no possibility of receiving a reply to any signal sent to the infalling observer. A somewhat tedious calculation (see appendix at end of article) shows this time, as measured on the ship, to be about one-fifth of an orbital period for large orbits. (This result is not unexpected since the orbital period of the ship is the only time scale in the problem.) Consequently, by the time Tony has completed just one orbit around a black hole, it is far too late to engage in two-way communication with Richard. At 10 Schwarzschild radii around a black hole of 10 solar masses, the critical time is far from infinite; rather, it is only about 0.005 second. The moral of the story is that a black hole—as they say—is truly black.

On another matter, one reader pointed out that our calculation of the tidal forces on the unfortunate astronaut in orbit around a 1-solar-mass neutron star assumed that he remained at a constant height of 10 kilometers. The reader rightly claimed that a situation more closely corresponding to Niven's story would be that the astronaut loops in only once and remains at 10 kilometers for only a brief amount of time, about one-tenth of a second, and thus might be able to withstand the large stresses involved. In principle this objection is correct; however, the tidal forces involved are so great that after even only one-tenth of a second, the astronaut's head is flying from his feet with a velocity of about 100 kilometers per second, producing an effect qualitatively similar to that previously estimated: pink applesauce splattered over the walls of the spacecraft.

IV. The Mills of God: Accretion Disks

Black holes have often been invoked as the powerhouses of the universe. They are, for instance, the currently favored mechanism to explain the astounding energy output of quasars, or QSOs, the strange objects that appear starlike in photographs but whose exact nature has remained a mystery since their discovery in 1963. Although most quasars are probably smaller than our solar system, they are substantially brighter than entire galaxies, often by factors of 10 or 100. At first sight, the black hole explanation for QSOs may seem strange. After all, a classical black hole is something that by definition does not emit radiation. How then, can any energy be gotten out of a black hole at all, let alone enough to power a quasar whose luminosity is 10^{13} times that of the sun?

In "Demythologizing" we talked at length about the Penrose process, a mechanism by which some of the rotational energy of a Kerr black hole could be extracted. In the most ideal situation, you could turn 120% of an object's mass into energy by $E = mc^2$. (The rotational energy of the hole added the extra 20% to that supplied by "energizing" the infalling object itself.) But we also mentioned that the Penrose process usually doesn't work very well. In a more realistic situation we could only expect about a 5% mass-conversion efficiency—still 10 times better than the 0.5% efficiency of hydrogen bombs. In this realistic scenario, the Penrose process has actually failed; you are *not* extracting the rotational energy of the hole. Instead, you have defaulted to a different energy extraction mechanism that, in fact, may be very important in real astrophysical situations. This is the process known as accretion.

Consider the situation in Fig. 1. An asteroid orbits around a black hole at a large distance labeled r_2. Let us say that we would like to attach a rocket to the asteroid and move it out to an infinite distance. We might think of detaching an asteroid from its orbit around the sun and transporting it to Alpha Centauri, which, for any reasonable purpose, can be considered infinitely distant. To do this requires energy. This is not surprising; the fact that the asteroid *is* in orbit means that it is bound to the central body. The energy needed to move the asteroid to infinity is appropriately termed the *binding energy* of the asteroid. It is the sum of the energy of the asteroid's motion (kinetic energy = $\frac{1}{2}mv^2$; m = mass of asteroid, v = velocity of asteroid) and the energy due to its distance from the black hole (potential energy = $-GMm/r_2$; M = mass hole). The binding energy for this asteroid is labeled BE_2 in Fig. 1.

Now, let us suppose that somehow the asteroid falls to the inner orbit, r_1. This orbit will have a binding energy BE_1. By the law of

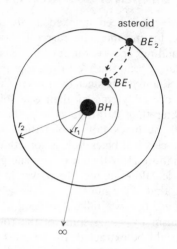

FIGURE 1 The energy required to move an asteroid from an orbit of any radius r to infinity is termed the binding energy of the asteroid. Suppose an asteroid is on orbit r_1 and is moved to r_2. This requires the input of an amount of energy $BE_2 - BE_1$, the difference in binding energies between the two orbits. Conversely, if the asteroid falls from r_2 to r_1, it must release an amount of energy $BE_2 - BE_1$. This is the energy available for accretion. To calculate the maximum energy available, one sets r_2 to infinity, which makes $BE_2 = 0$, and sets $r_1 = 6GM/c^2$, the last stable orbit. The calculation accompanying the figure then yields an upper limit of $(1/12)mc^2$, where m is the mass of the asteroid. Relativistic corrections lower this value slightly.

Binding energy of asteroid = kinetic energy + potential energy

$$BE = KE + PE.$$

For Newtonian orbits $\qquad PE = \dfrac{-GMm}{r}$

where M = mass hole, m = mass asteroid, r = radius orbit and

$$KE = \tfrac{1}{2}mv^2; \ v = \text{velocity}.$$

To be in circular orbit requires centrifugal force = gravitational force or

$$\frac{mv^2}{r} = \frac{GMm}{r^2}$$

so,

$$KE = \tfrac{1}{2}mv^2 = \frac{1}{2}\frac{GMm}{r} = -\tfrac{1}{2}PE.$$

Then, for a circular orbit,

$$BE = -\tfrac{1}{2}PE + PE = \tfrac{1}{2}PE = -\frac{1}{2}\frac{GMm}{r}.$$

Let radius of outer orbit $r_2 = \infty$, so $BE_2 = 0$.

Radius of black hole $= R = \dfrac{2GM}{c^2}$.

Let radius of inner orbit be last stable orbit around hole, $r_1 = 3R$. Binding energy available for accretion:

$$\cong BE_2 - BE_1$$

$$= 0 - \left(-\frac{1}{2}\frac{GMm}{r_1}\right)$$

$$= \tfrac{1}{2}GMm \cdot \frac{c^2}{6GM}$$

$$= \tfrac{1}{12}mc^2 \approx 8\% \text{ rest energy asteroid}$$

---

conservation of energy, the difference in these two energies, $BE_2 - BE_1$, must go somewhere. It will be the energy used to power the accretion disk. Before going that far, however, it may be useful to imagine a simple everyday example: dropping a rock to the ground. As we hold the rock motionless in our hands at a height h above the ground, the velocity v is zero, so there is no kinetic energy ($\tfrac{1}{2}mv^2 = 0$). But the rock does have a potential energy because it is positioned a finite distance from the center of the earth (potential energy $= -GM_{earth}\, m/(r_{earth} + h)$). In this case the binding energy $(KE + PE)$ is just the potential energy (PE). When the rock is dropped, it falls through the height h. Its final resting place is slightly closer to the center of the earth, so its binding energy has changed (by an amount known to physics students as mgh). What happens to the energy that was lost? On collision with the ground, the binding energy goes into increasing the motion of the atoms in the rock, and so the rock is heated. If the rock were heated enough, it would begin to glow and we would observe the binding energy emitted as visible radiation. This is exactly what happens in an accretion disk.

A black hole may be near enough another star to be able to siphon off matter from its companion by gravitational attraction. This matter, generally hydrogen gas, is expected to form a disk around the black hole and gradually spiral toward it. (A remarkable computer-generated photograph of a black hole accretion disk is shown in Fig. 2; further explanation is given in Fig. 3.) Each gas molecule can be viewed as a miniature asteroid in orbit around the hole. (You would *not* expect to find real asteroids in an accretion disk, nor the rainbows that appeared in Jonathan Post's *Omni* effort, April 1980.) Each orbit has a certain binding energy associated with it. As the molecule orbits the hole, it will generally undergo collisions with other molecules.

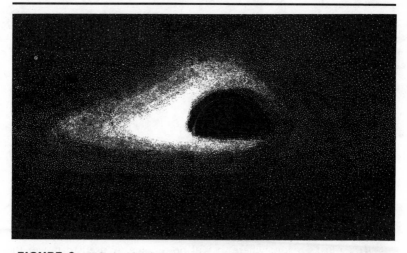

FIGURE 2 J.-P. Luminet's computer-generated photograph of a Schwarzschild black hole surrounded by an accretion disk. The observer holding the camera is situated at 10° above the equatorial plane of the disk, far from the hole (see Fig. 3). Although the black hole is not rotating, the disk is. The left wing of the disk is moving toward the camera, which causes the light to be Doppler-blueshifted. The right wing is Doppler-redshifted. The blueshift causes more photons to reach the photographic plate per unit time and each photon to have a higher energy than normal. The redshift causes the opposite effects. Thus the plate, which measures optical luminosity, records the left wing of the disk as being much brighter than the right wing.

The hump "above" the hole is actually a distorted view of the far side of the disk. The trajectories of photons emitted from that side are greatly deflected by the hole and into the camera. This will cause circles of constant brightness around the black hole to appear elongated. (Such circles are faintly visible in the photograph. See also Fig. 3.)

The ring of bright dots closest to the center of the photograph is composed of photons that have been gravitationally deflected by the hole so much that they may have circled the hole many times before escaping to the observer. The radius of this ring is *not* the Schwarzschild radius of the black hole, but approximately 2.6 Schwarzschild radii. Any photon initially passing the hole at a distance less than this value will fall into the hole and never be seen. Any photon passing at a slightly greater distance than 2.6 radii will circle the hole many times but finally escape to infinity. Any photon passing at 2.6 radii will spiral to an orbit of radius 1.5 Schwarzschild radii, the last orbit for photons. There the photon will orbit forever, never to be seen by a distant observer. Hence, any black hole illuminated by an external light source, like an accretion disk, will have an apparent radius of about 2.6 times the true Schwarzschild radius.

Finally, the dark band between this inner circle and the bright accretion disk itself is the onset of the region beneath 3 radii, where orbits of gas particles in the disk are unstable. These particles fall rapidly into the hole and do not have time to emit much radiation. Consequently, this area of the photograph appears dark.

For those interested in the technical details of generating the photograph, see *Astronomy and Astrophysics*, **75**, 228 (1979). We thank Dr. Luminet for permission to use this picture.

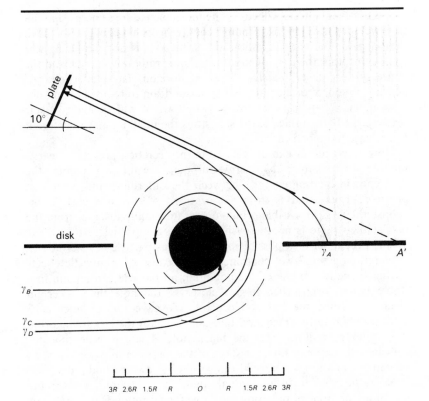

FIGURE 3 Guide to Fig. 2. The photographic plate is located at 10° above the plane of the disk. The black hole has radius R. The inner edge of the accretion disk is at $3R$, the last stable orbit for massive particles. Photons such as the one labeled γ_A will be emitted by the disk and deflected toward the plate. The observer's eye will project back along the trajectory and assume γ_A was emitted from point A'. This will cause the rear part of the disk to appear elongated, producing the hump in Fig. 2. Photon γ_B approaches the hole at a distance less than the "critical impact parameter" $\cong 2.5980762122R$. It will be captured by the hole. The photon γ_C approaches the hole at exactly the critical impact parameter and continues to orbit the hole forever at a radius of $1.5R$. The photon γ_D passes the hole slightly outside the critical impact parameter. It may orbit the hole many times before escaping outward, where it may be observed. The observer will trace back the trajectory in a straight line (not shown) and will assume the photon was emitted from the hole at a radius slightly greater than $2.6R$. Hence, photons such as γ_D are responsible for the bright inner ring in Luminet's photograph.

For a technical discussion of photon orbits around black holes, see *Gravitation*, by Charles Misner, Kip Thorne and John Wheeler (W. H. Freeman & Co., San Francisco, 1973), chap. 25.

These frictional effects will cause the molecules to lose energy and fall to lower orbits. This lost binding energy heats up the gas so much that large amounts of radiation are emitted. It is not the black hole itself, but rather the gas falling into it, that is radiating energy, and the entire process occurs *outside* the event horizon. The black hole just acts as the source of the potential and as a drain to remove the matter after it has fallen in. The process won't work with a neutron star because the hard surface of the star stops the infalling matter before it can release too much energy.

How much energy can be released in the accretion process? Clearly, in order to maximize the energy output, we want to maximize the difference in binding energies between the outermost and innermost possible orbits—that is, to maximize $BE_2 - BE_1$ in Fig. 1. If we choose the largest possible orbit—one at an infinite distance from the black hole—there is no energy binding the gas particle, so $BE_2 = 0$. What is the innermost possible orbit for a gas particle or asteroid? In "Demythologizing" we mentioned that because of relativistic effects on the shape of the black hole potential, no particle can remain in a stable bound orbit when it gets closer to the hole than a certain distance, termed the last stable orbit. We gave this distance as 3 Schwarzschild radii. Once a particle falls beneath the last stable orbit, it is quickly swallowed by the black hole. There is little time for collisions to occur in this last stage of the gas's spiral to oblivion, so very little radiation is released from within 3 Schwarzschild radii.

Now the answer to the initial question is straightforward: the maximum amount of radiation available for accretion is $0 - BE_{lso}$, or negative the binding energy of the last stable orbit. The simple calculation accompanying Fig. 1 shows that one-twelfth the rest mass of the infalling particle is converted to energy, roughly 8%. This calculation was done for Newtonian orbits. In practice, many effects, including those of relativity, tend to reduce this value, and the previously mentioned 5% is probably closer to the truth.

Those who view black holes as magical objects lying outside the normal realm of physics may be disappointed by the pedestrian nature of the above explanation. As pedestrian as it may be, however, black hole accretion is the only mechanism yet proposed that seems to be able to explain quasars. Yet, one should not be tempted to think that the subject of accretion is entirely simpleminded. Consider, for instance, spherically symmetric accretion. If, as some theorists propose, accretion disks around black holes are very thick objects, there may be situations in which the gas is falling into the hole from almost all directions. If the angular momentum of the gas is small, this situation would be idealized by gas falling radially into the hole— spherically symmetric accretion. Any radiation released in such a process would have to pass out through the infalling gas and thus interact with it. The problem of black hole accretion is technically very difficult and it is fair to say that there are fundamental disagreements between various researchers in the field. A solution to this riddle would be of the greatest theoretical interest, but we would not be surprised if it is several years in forthcoming.

V. The Princess and the Pea: Black Hole Telegraphs

Speaking of accretion, an amusing suggestion for communication between supercultures appeared in Jonathan Post's previously mentioned *Omni* article. He proposed that we put asteroids into orbit around a black hole of unspecified mass. To send a telegram to a friendly neighboring supercivilization, we detonate hydrogen bombs to stop the asteroids in their paths. The asteroids will then fall into the black hole and release some energy. Now, you might well think that the energy extraction mechanism Post proposes is to use the binding energy of the asteroids in the accretion process just discussed. No, the idea is far more elegant. When an asteroid falls into the hole, a certain portion of its mass will also be converted into gravitational radiation and be emitted as a burst of gravitational waves. By stopping many asteroids at calculated intervals and allowing them to fall into the hole, these gravity-wave bursts could be used as a sort of intergalactic Morse code (see Fig. 4).

As mentioned in "Demythologizing," gravitational waves are very similar to ordinary radio waves. Relativity predicts that, like radio waves, gravity waves will be emitted by any object that undergoes acceleration. Unlike radio or light waves, however, gravity waves pass easily through interstellar and intergalactic material, making their attractiveness for use as a telegraph easy to see. Unfortunately, their strength is almost always ridiculously small, as the following calculation dramatically illustrates. We present this calculation, not to burden you with unnecessary details, but to show that straightforward reasoning often allows us to decide immediately on the feasibility of a proposal.

Since the exact numbers won't matter very much, let us assume we are using 10-megaton hydrogen bombs to stop the asteroids. Now, the Hiroshima bomb had a yield of about 10 kilotons and required roughly 100 kg of uranium, so a 10-megaton bomb should be the equivalent of 10^5 kg (100 metric tons) of uranium. The efficiency of hydrogen bombs, on the other hand, is only about 1% (a liberal estimate). Consequently, we are only going to convert 10^3 kg of uranium into energy. Using $E = mc^2$, where $c = 3 \times 10^8$ meters per second, we find that the 10-megaton bomb releases roughly 10^{20}

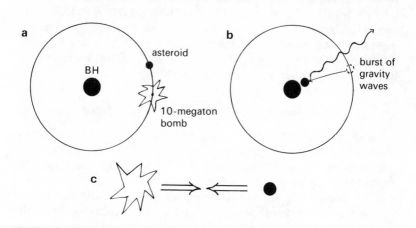

Momentum bomb radiation = momentum asteroid

$$\frac{E}{c} \approx mv$$

$$\frac{10^{20}}{3 \times 10^8} \approx m10^4$$

$$m \approx 10^8 \text{ kg}$$

But efficiency of gravity-wave production $\sim \dfrac{m}{M} \sim 10^{-22}$.

So amount of asteroid released as gravitational radiation

$$\sim m \cdot \frac{m}{M} \cdot c^2 \sim 10^8 10^{-22} 10^{17} \sim 10^3 \text{ Joules}$$

$$= 100\text{-watt lightbulb burning for 10 seconds}$$

FIGURE 4 A hypothetical "black hole telegraph." In **(a)** a hydrogen bomb explodes to stop an asteroid that has been in orbit around a black hole. Once stopped, the asteroid can fall into the hole as shown in **(b)** and release a burst of gravitational radiation. Unfortunately, equating momenta in **(c)** shows that only a 10^8-kg asteroid can be stopped, and the gravitational energy produced is no greater than the energy used by a 100-watt lightbulb shining for 10 seconds.

(The necessity of the hydrogen bomb illustrates a significant point overlooked by many science-fiction writers—namely, conservation of angular momentum. How many stories or movies have involved throwing a dangerous missile or garbage from a spaceship into the sun in order to save the crew and world? This simply can't be done, because the missile [asteroid] has an angular momentum corresponding to the particular orbit it is on. One must first get rid of this momentum [e.g., with a hydrogen bomb]; then the missile can fall into the sun, not before. One wonders why the same writers have never asked themselves why the earth didn't fall into the sun long ago.)

joules, enough energy to power the United States for three or four months. (This is obviously a vast overestimate—hydrogen bombs do not weigh 100 tons. A 10-megaton bomb actually releases 4×10^{16} joules.)

This energy must be used to stop an asteroid cold, so that it can then fall out of orbit and into the black hole. In more scientific terms, the momentum of the bomb's radiation must equal the momentum of the asteroid (see Fig. 4). The momentum of any massive body is given by the formula $p = mv$, where m is its mass and v its velocity. The momentum inherent in the radiation from the bomb is given simply by $p_r = E/c$. If we assume that the asteroid has a velocity characteristic of all planetary bodies, about 10^4 meters per second, equating the two momenta shows that the asteroid can have a mass of at most 10^8 kg (100,000 metric tons) if it is to be stopped. This is in the mass range of battleships.

As the asteroid falls into the hole, a certain fraction of its mass will be converted into gravitational radiation. The efficiency of this process is not very high; in fact, a generous upper limit is *efficiency* $\sim m/M$, where m is again the asteroid mass and M is the mass of the black hole. The smaller the black hole, the higher the efficiency. Thus, let us choose the smallest feasible black hole, about 1 solar mass or 10^{30} kg. For a 10^8-kg asteroid, the efficiency in generating gravitational radiation is therefore only 10^{-22}, which means that the amount of the 10^8 kg released as gravitational radiation is $10^{-22}10^8 = 10^{-14}$ kg. Using $E = mc^2$ again, this amounts to a total of 1000 joules per asteroid, the equivalent of burning a 100-watt lightbulb for 10 seconds—not a very bright beacon to signal a friend on the other side of the universe. As a matter of fact, the bomb explosion itself released 10^{17} times as much energy. Unless this supercivilization is superstupid, the rationale for gravity-wave telegraphs remains, at best, opaque.

On a more serious level, the incredibly small result of this calculation gives some good intimation of the difficulty of detecting gravitational radiation. Because the waves are so weak, extraordinarily sensitive detectors are required to observe them. The famous "Weber bars," the gravity-wave detectors first constructed by Joseph Weber, were large, 1000-kg, aluminum cylinders designed to vibrate when a gravity wave passed by. Such cylinders could not sense the previously discussed asteroid falling into the black hole from a distance greater than about 10 meters, let alone across the galaxy. Present detectors could probably pick up a 10-solar-mass star collapsing into a black hole in our galaxy, and in the next decade we hope to have detectors that could sense a supernova in the Virgo Cluster. (A more realistic view of what a gravity wave propagating from a collapsing star might look like is shown in Fig. 5.) Such detectors might be constructed of a 1000-kg monocrystal of quartz or sapphire, cooled to a thousandth of a degree above absolute zero in order to reduce the thermal vibrations inherent in any solid—vibrations that would tend to mask vibrations caused by gravity waves. These detectors will have to be sensitive to vibrations of less than 10^{-19} cm over durations of one-tenth of a second. In light of the above discussion, the necessity for such refinements is not entirely surprising, nor is the fact that gravity waves have not yet been discovered despite the persistent efforts of many researchers over the past 15 years.

VI. Lobsters and Grand Pianos:
Why the Universe Is Not a Black Hole

One of the questions that relativists are most often asked is, "If the universe is closed, are we then living inside a giant black hole?" The

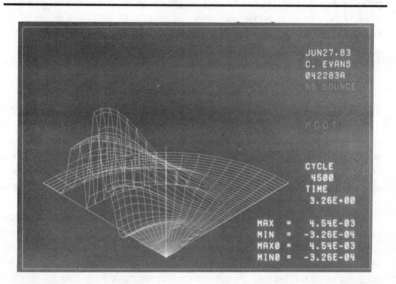

FIGURE 5 A more realistic view of the distortion of spacetime caused by a gravitational wave. This particular wave is propagating outward from a collapsing neutron star. The neutron star is taken to be at the center of the picture with its north pole pointing toward the word "cycle." The leftmost axis lies on the equator. Only one-quarter of spacetime is shown; to generate the full spacetime, the figure should be rotated around the polar axis. (We would like to thank Charles Evans for supplying this figure which he produced on the Cray-1 computer at Lawrence Livermore Laboratory.)

question is so popular that Isaac Asimov devoted a section of his book *The Collapsing Universe* to it and—unfortunately—arrived at exactly the wrong conclusion. But he is not alone—many physicists have the same misconception. As indicated by the above title, the universe is as much like a black hole as a lobster is like a grand piano; they both obey the same laws of physics, and there the similarity ends.

The origin of the "black hole universe" legend remains lost to prehistory, but part of the confusion may have arisen because the equation giving the radius of the observable universe as a function of its matter content looks suspiciously like the equation for the Schwarzschild radius of a black hole:

$$R_U \approx 2GM_U/c^2$$

and

$$R_{BH} = 2GM_{BH}/c^2.$$

The twiddle symbol ($\approx$) means the relationship is approximate. In the case of an open universe (one that will never recollapse under gravitational attraction) or a closed universe (one that will eventually recollapse), the approximation may be good or bad depending on the age of the universe. For a "marginally bound" universe (one on the borderline between being open and closed), the relationship turns out to be always exact—the binding energy of such a universe is zero.

The reason the above two equations are so similar is that the dynamics of both the universe and black holes is governed by one and the same theory of gravity. But too much can be made of such a resemblance; if we divided the size of a lobster by its mass, the result would probably be reasonably close to the R/M value for a grand piano, too.

Although the two formulas look similar, they nevertheless mean very different things. In the first case, M_U consists of just the total matter content of the universe: dust, galaxies, stars, neutrinos, photons. . . . It does *not* include the binding energy of the universe. By contrast, M_{BH} encompasses the entire energy content of the black hole, binding energy of the original star included. Now the binding energy of a star—the energy that holds it together—can be a very large percentage of its total energy. As a result, the M's in the two formulas stand for two different quantities and may have substantially different values.

Furthermore, although the universe is expanding, a black hole is not—the Schwarzschild radius of a black hole is a constant. If you called R_U the "Schwarzschild radius of the universe," you would be in the uncomfortable position of having a constant that continually increases.

There are other, more important reasons why the universe cannot be a black hole. There is the purely technical point that the mathematics governing the description of a black hole requires that it be embedded into—that is, be part of—an outside universe. In mathematical terms, the black hole requires boundary conditions to match it with the rest of the universe. On the other hand, the universe itself *is* everything and therefore cannot be embedded in something larger. There are no boundary conditions matching the entire universe with something else. This means, for instance, that things can fall on you from the outside of a black hole but cannot fall on you from the outside of the universe.

You may object that perhaps everything astronomers have ever seen is only part of a black hole residing in a vastly larger universe. To see why this is not the case, recall that Richard knew he was falling toward the center of a black hole. Black holes define a direction. Falling into a hole, observers experience an asymmetry as to what they may see by looking in various directions. Astronomers in the real universe see none of the precisely defined asymmetries that the infalling observer experiences while plummeting toward the singularity. So the question of whether the universe is a black hole is also given an observational answer: it isn't.

VII. A Fish's-Eye View of the Outside World: How the Universe *Does* Appear to an Infalling Observer

We have already discussed at length many objects falling into black holes: gas, asteroids and astronauts. Moreover, the discussion has centered on the view of an external observer—Tony watching Richard's plunge and an accretion disk as viewed from far away. It was mentioned previously that an infalling observer sees asymmetries in his view of the outside world, but we have not yet actually described in detail what such an adventurer really would see.

This interesting topic has also been frequently misreported. Most notorious is the belief—even to be found in some undergraduate textbooks—that as an astronaut falls to the singularity, he sees not only his life pass before his eyes, but the entire future history of the universe, as well! In a probably related vein, Jonathan Post claimed in his article that an astronaut would see life in a space station directly above him speeded up to undefined frenzy. Both statements are incorrect.

As to the first statement, recall yet again Richard's plunge to the singularity. In Section III we showed that a last photon from the searchlight reached him as he crossed the event horizon. This defined a last time that two-way communication was possible. Similarly, a last photon from ship or star reaches Richard as he hits the singularity. The corresponding time interval as measured on the ship between Richard's jump and the emission of this last photon will be the amount the universe has aged. The time is comparable to those already discussed; once beneath the event horizon of a solar-mass black hole, Richard will see the universe grow older by about 0.000015 seconds. For a billion-solar-mass hole this time is increased by a factor of a billion to roughly 4 hours.

As to the second statement, the appearance of a star or space station to the infalling astronaut depends on its position in the sky. Light from an object *directly* above the astronaut is actually red-shifted, which corresponds to the slowing down of clocks. Life in the space station would not appear frenzied, but rather leisurely.

In any case, we can calculate and describe what an astronaut sees as he falls into a black hole. There is no need to resort to metaphysics. As a preliminary, recall that the appearance of the accretion disk in Fig. 2 is largely determined by the fact that the trajectories of photons are deflected in the presence of gravitational fields. This deflection, of course, occurs around any massive body. Jupiter deflects light by about 0.02 seconds of arc, and the famous 1919 test of general relativity showed that light passing the sun was focused slightly inward, by the small amount of 1.75 seconds of arc. As we have already seen, the large gravitational fields of black holes can make this effect enormous. In the captions to Figs. 1 and 3, it was pointed out that any light ray initially passing within 2.6R of the hole will be captured; any ray passing at 2.6R will orbit the hole forever at a radius of 1.5R; and any ray passing slightly outside 2.6R will circle the hole a few times and then escape to infinity. Because of these various paths by light rays around a black hole from, say, a star to an observer, any star will produce a number of images visible to that observer (see Fig. 6).

The first, or primary image, is that which we would normally see without the presence of a black hole. In the presence of a black hole, the primary image is a direct image caused by light that passes by the hole, is somewhat deflected and then travels into the eye of the beholder. The star will appear displaced slightly outward from its normal position. On the other side of the hole will be a second image caused by a ray passing the hole on that side, deflected the proper amount to reach the observer's eye. A third image is caused by the ray that circles the hole once and reaches the eye. In principle there are an

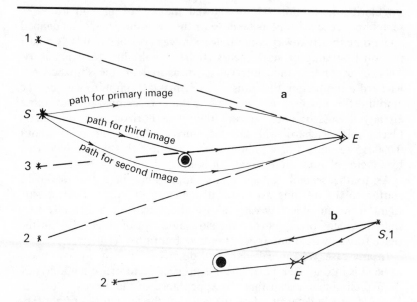

FIGURE 6 In diagram **a**, the star (S) sends light to the eye (E) via many paths around the black hole. The primary image (1) is caused by the most direct path to the eye and causes the star to appear shifted outward from the black hole. The second image (2) is caused by a ray passing on the other side of the hole; the third (3) by a ray wrapped once around the hole. If the star is behind the observer, as in **b**, the primary image will be the true star position, while the secondary image will appear on the opposite side of the sky. He can thus see the same star by turning his back on it!

infinite number of such images. All lie on a line through the black hole and on alternating sides of the hole. The higher-order images, however, are extremely dim (each order is about 20 times dimmer than the previous one) because there is very little light aimed from the star exactly along the paths necessary to form one of these higher-order images. Consequently, only the first two or three are generally visible, even in the presence of black holes.

There is another effect that greatly influences how an observer traveling at high velocities views the universe. This effect has nothing to do with black holes and is termed *aberration*. The standard explanation will serve our purposes: On a rainy day, a stationary observer will see the rain falling straight down, but someone driving in a car will see the rain falling at an angle. The observer in the car does not feel himself moving, but the rain now has not only a downward component of velocity, but a horizontal one as well. Thus, it falls at an angle. The same phenomenon occurs with starlight and has been known for several centuries. The observations of star positions made by someone moving at the speed of light will be pronouncedly different from those made by a stationary observer.

Both aberration and the bending of light by gravitational fields will be responsible for the spectacular light show to come. With this background, let us now imagine a journey into a billion-solar-mass black hole. This hole is about the size of our solar system. Our sacrificial astronaut, Richard, will begin his journey at one-tenth the velocity of light at a distance of 10,000 Schwarzschild radii from the black hole. We will indicate his progress by giving both the time remaining before he finally reaches the singularity, and his distance from the singularity as measured in Schwarzschild radii.

The beginning of the journey, 30 years before the end. At this point, the black hole is rather unimpressive. There is a small region (about 1° across—*i.e.,* twice the size of the moon) in which the star patterns look slightly distorted. A careful examination of the stars shows that a few of those nearest the hole have second images on the opposite side of the black hole. Had this not been pointed out to him, Richard probably would have missed the black hole entirely.

Ten days before the end, at 16R. The image has now grown immensely. There is now a pure dark patch ahead with a radius of about 10°. The original star images that lay near the direction of the black hole have been pushed away from their original positions by about 15°. Furthermore, between the dark patch itself and these images lies a band of second images of each of the stars we can see ahead of us. Looking near the darkness with the aid of a telescope, we can even see faint second images of stars which actually lie behind us (Fig. 6)!

In Frames A and B in Figure 7 we have a diagram of Orion both with and without the black hole present. The view corresponds approximately to what Richard would see ten days before the end. The shape of this familiar constellation has become severely distorted, its second image even more so. The arrows numbered 1 point to the primary images of Betelgeuse, Rigel and the other main stars of Orion. The 2's point to the second images.

Four hours before the end. Richard is now at the Schwarzschild radius and is thus traveling near the speed of light. Aberration effects will now be extremely important. Anything he sees from this moment on will be a secret taken to the grave because he can no longer send any information out to us. Nonetheless, we will still follow him in our imagination. Although Richard is now inside the black hole, the sky in front of him does not appear entirely dark. Because he is traveling

Frame A

Frame B

FIGURE 7 Two frames from Bill Unruh's computer-generated film of an astronaut in orbit around a 1-billion-solar-mass black hole. The astronaut is 18 Schwarzschild radii from the hole. Frame A shows an unobstructed view of the sky around the constellation Orion. Orion itself is clearly visible in the center. To the upper left is the constellation Auriga and far in the upper right are the Pleiades. Some of the more familiar stars are labeled in the key. Frame B shows the same section of sky with the black hole in front of the astronaut. The gravitational field has deflected light paths so much that Orion is barely discernible. Furthermore, both primary and secondary images are now visible, making the pattern even more confusing. Some of these images are labeled in the key.

Legend: Betelgeuse = B;   Rigel = R;   Belt   stars = Bl;   Aldebaran = A; Pleiades = P; Sirius = S; Primary image = 1; Secondary image = 2.

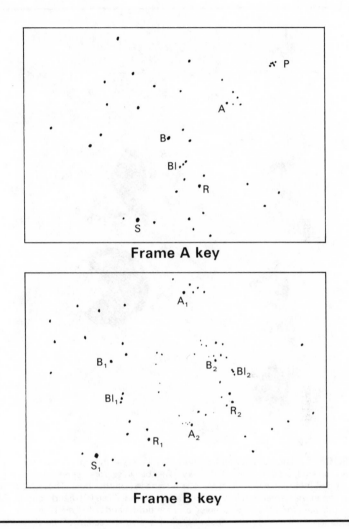

Frame A key

Frame B key

near the speed of light, aberration causes light rays to arrive at his eye at extreme angles. In fact, only the patch immediately in front of him, subtending an angle with radius 45°, appears black. This is the apparent size of the black hole—a substantial fraction of the forward sky.

Behind him, Richard sees the stars there growing dim and spreading out. They are moving around to meet the advancing edge of the black hole. This is again an aberration effect (see Fig. 8). But there is a more noticeable feature of the sky. Richard can now see second

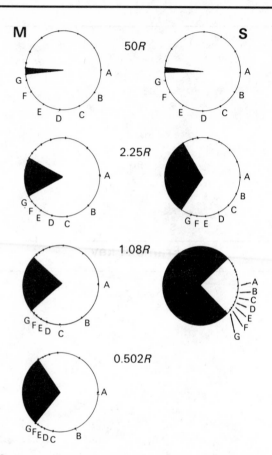

FIGURE 8   The sky near a black hole. The letters A through G represent stars evenly spaced around half of the sky. The black sectors represent the angle subtended by the black hole, at various distances from the hole. These distances are given in the center in terms of Schwarzschild radii. The left-hand column (M) gives the view for a moving observer, and the right-hand column (S) the view for a stationary observer. We see that at 50R only a small section of the sky is dark and the star positions are undisturbed, except for those stars nearest the hole. At about 1R, the stationary observer sees most of the sky black. The stars are squeezed around behind him because of the bending of light around the black hole. On the other hand, the moving observer is now traveling at almost the speed of light. To him, only a 90° angle (45° in radius) in front of him is dark, and the stars are moving toward the hole. Even at about $\frac{1}{2}R$, most of the sky is still light. We also see that the star positions are in general more distorted for the moving observer than for the stationary observer, showing that it is the aberration effect that is mostly responsible for changing the star positions. (This figure is based on the work of M. Sikora, courtesy M. Abramowicz.)

images of all the stars in the sky surrounding the black hole. They are all squeezed into a band about 5° wide. The second images are now brighter than were the original stars, because of the blueshifting of light falling into the hole. Surrounding the ring of second images are the still brighter first images. Three-quarters of these images are concentrated in a band 20° wide lying immediately outside the band of secondary images. The rings run completely around the sky, from above Richard's head to beneath his feet. The band of light caused by both the primary and secondary images now shines with a brightness ten times that of the normal night sky.

Four minutes to oblivion, at 1/15 R. The apparent size of the black hole is now 75 degrees in radius—almost the entire forward sky. Behind Richard stars are dimming and rushing forward. Only 20% of the stars are left in the sky behind him. In a 10°-wide band surrounding the outer edges of the black hole, not only second, but also third and some fourth images are now visible. This band running around the sky now glows 1,000 times brighter than the night sky.

The final seconds. The sky everywhere but in that rapidly thinning band is dark, while the band itself glows ever brighter. At 3 seconds before eternity, it shines brighter than the moon. New stars are rapidly appearing along the inner edge of the shrinking band as higher and higher-order images become visible from light wrapped many times around the black hole. The stars of the universe seem to brighten and multiply as they are compressed into a thinner and thinner ring. Only in the last tenth of a second do the tidal forces become strong enough to end Richard's journey and his view of that awesome ring bisecting the sky.

As we have repeatedly stressed in these articles, there is no need to exaggerate the properties of black holes. Before turning our attention to other matters, let us go back to the accretion disk. Some time was spent in describing what the disk looks like to a distant observer. In the spirit of our just-completed journey into the black hole, it is now appropriate to give an idea of the appearance of the universe-at-large to a cosmonaut standing on the accretion disk itself. This we do in Fig. 9. Because you have by this time been immersed in bending light rays, the explanation in the figure caption should suffice.

VIII. Son of Kong: Tipler Machines, Naked Singularities, White Holes

Our friend and colleague Frank Tipler, author of the last article in this book, would probably not object to being termed controversial.

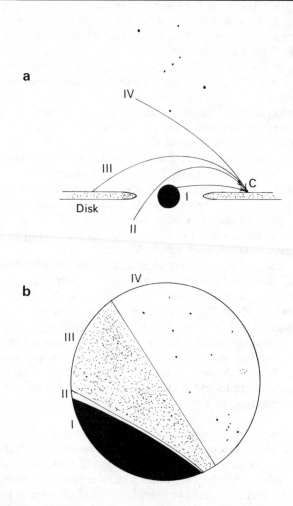

FIGURE 9 Polnarev and Turchaninov's calculation of the celestial sphere for a cosmonaut standing on the accretion disk surrounding a rotating black hole. The cosmonaut, at position C, receives no rays from the hole (I), but does receive rays from beneath him via paths II; from the disk itself via paths III; and from the night sky itself via IV. Figure B shows the cosmonaut's view of the sky if he is standing at $6R$ from the singularity. You should imagine yourself lying on your back staring straight up, the black hole on your left. If, in the absence of a black hole, Orion were at the zenith, then in the presence of a black hole, Orion would appear approximately where shown to the cosmonaut on the disk. It would also be considerably compressed, as indicated (although this effect was not drawn to scale). (See *Acta Astronomica*, **29**, 81 [1979].)

His paper, "Rotating Cylinders and the Possibility of Global Causality Violation" has certainly generated enough controversy to please even him. After the paper appeared in a 1974 issue of *The Physical Review*, fringe elements immediately heralded it as a breakthrough: for the first time in a professional journal the words "time machine" appeared. Not too long ago Poul Anderson used the idea in his novel *The Avatar*, and even more recently the punk rock magazine *Wet* actually termed Tipler the inventor of the time machine. (*Wet* will be pleased to know Frank enjoyed the tongue-in-cheek humor thoroughly.) *Omni* magazine devoted substantial attention to Tipler machines in an article by Robert Forward (May 1980).

It is a pity. Most of the hoopla has unfortunately arisen from telepaths with faulty reception who have incorrectly divined the contents of the paper. Certainly they have not read the original (admittedly, it is for the specialist). In that paper, Tipler pointed out that an astronaut orbiting a long cylinder rotating at one-half the speed of light could travel into his own past. This is true. (Such cylinders may be thought of as progeny of rotating black holes, whose lack of time-travel properties was discussed in "Demythologizing.") What everyone fails to mention, or discards as irrelevant, is that Tipler did his calculation for an *infinitely* long cylinder, a mathematical abstraction used to simplify a difficult calculation. He then *speculated* that the time-travel effect *might* take place around a very long finite cylinder, which would be viewed as "approximately infinite." *But he did no such calculation.*

Most relativists are convinced that time travel *could not* work using finite cylinders for the following reason: An infinite cylinder, by its very nature, requires nothing to hold it up. Tipler's cylinder had no internal pressure at all. The situation is akin to that of a star collapsing into a black hole, but because the cylinder is infinitely long, the collapse never gets anywhere; the cylinder always remains infinite in length. A finite cylinder, by contrast, requires the material of which it is made to exert a sufficient pressure to hold up the ends, ensuring that it doesn't noticeably collapse. Again, this is analogous to the pressure caused by the nuclear burning in a star that keeps the star from collapsing into a black hole. Now, there is an absolute maximum amount of pressure that any ordinary matter can exert. This maximum pressure corresponds to the condition that the matter have "positive energy," i.e., that the matter be gravitationally attractive—as all known matter is. The only way this maximum pressure could be exceeded would be if the matter were allowed to have "negative energy"—a condition that might loosely be thought of as gravitationally repulsive. Although to our knowledge the calculation has not been done, the positive energy condition appears to rule out any cylinders long enough to be regarded as "approximately infinite" for the purpose of time travel because the pressure required to hold up the ends would in turn require the cylinders to have negative energy.

In his *Omni* article, Robert Forward dismisses the positive energy requirement as an engineering detail, pointing out that such things as transistors did not exist one century ago. There is a substantial difference, however, between purifying silicon—a process that obeys every law known to physics—and postulating the existence of a substance that may in fact violate every known law of physics. We expect that it will be somewhat longer than a century before an engineer produces a sample of negative mass.

Just as black holes and Tipler machines have been proposed as star gates and time machines, so also have the more bizarre objects called naked singularities. In order to discuss these "things" (an appropriate label, since no one really knows much about them), let us backtrack to Kerr black holes. In 1963 the brilliant relativist and championship bridge player Roy Kerr discovered the mathematical description of spacetime around a rotating body, now called the Kerr metric. Imagine a spinning star shrinking to black hole size, rotating faster and faster as it does so. If the surface velocity of the star remains less than a certain critical value that is, in some sense, one-half the speed of light on the equator, the Kerr metric will always describe a black hole. But if the velocity of the event horizon exceeds this critical value, the event horizon vanishes and the singularity in the interior of the Kerr metric becomes visible to the outside world—hence, "naked singularity."

The interior of the Kerr metric contains closed, timelike curves (the technical term for orbits that allow an astronaut to travel into his own past). Presumably, the possibility of entering a Kerr metric and being able to get out again—because there is no event horizon—is what gave people the idea of using naked singularities as time machines. The objections are formidable. First of all, no one has ever figured out a way to make a naked singularity; every theoretical attempt ever made to spin up the Kerr metric to the critical value has failed. (Usually the process involves shooting objects at the Kerr metric in such a way as to increase its angular momentum.) In addition, the first computer simulations that actually collapse rotating Kerrs (recently made by Nakamura and Sato) fail to form naked singularities when the critical value is reached. Nature just doesn't seem to allow the beasts. In fact, Roger Penrose has postulated a fundamental law, the Cosmic Censorship Hypothesis, which, if true, excludes them altogether. Furthermore, assuming the positive energy condition mentioned earlier, it has already been proven in several specific cases that naked singularities cannot exist. Finally, if naked singularities *did* exist, their behavior would be so violently acausal that to predict their use as time machines would itself be an unpredictable prediction.

We are frequently asked, "What about white holes?," and sometimes fall into the temptation to answer, "What about them?" At least one popular book has been written on the subject, John Gribbin's *White Holes*, and rumor has them to be either time-reversed black holes or the far end of a wormhole—a gusher that spews forth matter that has fallen into a black hole at the wormhole's other end. Someone once proposed white holes as an explanation for quasars, but we don't think anyone has identified them with the quantum black holes of the next section. In any case, if white holes are supposed to be the other exit of a wormhole, no matter will get through because of the wormhole instabilities already described at length in "Demythologizing." If all this sounds rather vague, it is.

The difficulty in taking white holes seriously is best illuminated by a consideration of their theoretical origins. For the 1971 Texas Symposium on Relativistic Astrophysics, Roger Penrose reports that he was scheduled to give a talk entitled "Black Holes and White Holes." Not knowing what a white hole was, he prudently decided it would be wise to find out before the lecture. Upon arriving at the symposium, he therefore set about questioning various participants, only to discover that a white hole was either a time-reversed black hole, or a naked singularity or maybe even a black hole that was having its rotational energy extracted via the Penrose process. During the lecture he "invented" white holes by projecting a spacetime diagram of a black hole upside down. Using a 10-foot-long pointer so that he could remain at a safe distance from the beast, he then made a joke that neither he nor anyone else can now remember. And thus the world witnessed the birth of white holes.

IX. The Pit and the Particle: Quantum Black Holes

In "Demythologizing" we promised to discuss quantum black holes in a future article. We will now fulfill that promise. Unfortunately, no one has a clear picture yet of what happens near a black hole when quantum mechanics is taken into account. Popular descriptions, such as some sort of "tunneling" of particles from inside the black hole to the outside world, have been given, but we feel that none of these capture the reality of the quantum process near the black hole. Even in quantum mechanics terms, nothing actually escapes the hole. What seems to happen is that in the region near the horizon out to about 2 Schwarzschild radii, the quantum vacuum of spacetime experiences some sort of instability. So-called "virtual" particles (a particle and anti-particle) come into being. Virtual particles do not, normally, survive long enough to be measured (for reasons that will be discussed later), but with a black hole present, one may be swallowed by the hole while another flies off to infinity or to an observer. To make up for the energy required to create the observed particle, it seems that the infalling particle must carry negative energy into the black hole from the region in which the virtual particles were created. (Although negative energies are not allowed in classical physics, a very slight quantity of negative energy can evidently be accumulated in some regions if quantum mechanics is taken into account.)

This was the remarkable discovery of Stephen Hawking in 1974, for which he is justly famous. The emission of particles "by" black holes is called Hawking radiation or Hawking evaporation in his honor. Because the emission of particles must extract energy from the black hole, under the Hawking process the mass of the black hole actually decreases and eventually evaporates away altogether. (In terms of the previous discussion, one can say that the mass of the hole decreases by the addition of negative energy from the infalling particles.) It should be stressed that no one has a coherent physical picture of the process. What exists is a mathematical formalism that is very difficult to interpret physically. Although several pictures, such as our own, are often encountered to explain Hawking radiation, none of them is very satisfactory upon closer scrutiny. While everyone agrees that Hawking radiation is too pretty not to exist, there is still some debate about the validity of the mathematical methods employed and the exact nature of particle emission.

Because quantum mechanical effects usually are not important on scales larger than atoms, solar-mass black holes would not be expected to emit substantial Hawking radiation. One way to measure the strength of the radiation is by the temperature associated with it. A purely classical black hole does not emit radiation and thus has a temperature of absolute zero. A solar-mass black hole has a Hawking temperature of about 10^{-7} degrees absolute, a small number on any temperature scale. The smaller the black hole, the higher the temperature; this is in accord with our notion that quantum effects will become important for small objects. A 1-gram black hole, which has a Schwarzschild radius of about 10^{-28} cm, is about 10^{15} times smaller than a proton. While gravitational effects are still overriding, those of quantum mechanics are not far behind: this black hole has a temperature of 10^{26} °K, a very large number on any temperature scale.

Why talk about such small holes at all? Clearly, they would not form from the gravitational collapse of massive stars. We are forced to admit that such holes might not exist; there is no theoretical argument that requires their existence, and certainly no such holes have ever been detected. (There is, of course, White's totalitarian principle: "If something is not expressly forbidden, then it is compulsory.") Nonetheless, in the first blink of God's eye, the universe would have been dense enough to form such objects, called primordial black holes. More scientifically, the density of the universe would be greater than that of these 1-gram black holes for 10^{-38} seconds after the big bang. A small fluctuation in the initial chaos might snap some regions of the fluid into black holes. One-gram primordial black holes would emit particles via Hawking radiation at such a rate that they would disappear completely in about 10^{-26} seconds—10^{12} blinks of His eye—and release 10^{14} joules of energy. This is comparable to the yield of a small hydrogen bomb.

We have just described a very explosive event. Because the temperature and hence the rate of evaporation goes up as the mass of the hole goes down, a black hole loses energy at ever increasing rates, ending its life in a catastrophic burst of energy. Primordial black holes of about 10^{15} grams, about the mass of an asteroid but with the size of a proton, would be exploding during the present epoch of the universe and releasing large numbers of X rays into the cosmos. Now, astronomers observe there to be roughly 10^{-36} gm/cc worth of X rays in the universe. If *all* this is from black holes whose *entire* masses are converted into X rays, then there is at maximum 10^{-36} gm/cc worth of 10^{15}-gm black holes in the universe. Because the average matter density of hydrogen and other elements in the cosmos is about 10^{-30} gm/cc, this implies that 10^{15}-gm black holes can make up at most one-millionth the mass of the universe, about the ratio of the earth's mass to that of the sun. Further refinements in the calculations, by Page and Hawking among others, have made this upper limit even smaller.

Along the same lines, we can consider the effects of 10^{10}-gm black holes on the nucleosynthesis of the light elements, which occurred 3 minutes after the big bang. At that time, the temperature of the primordial soup dropped to about 1 billion degrees. (The central temperature of the sun is about 10 million degrees.) Neutrons and protons, previously too hot and traveling too fast to stick together, were now moving slowly enough so that nuclear forces could bind them into deuterium. Deuterium, however, likes to burn into helium and thus, in a nucleosynthesis chain not unlike that of the sun, the universe recycled the free neutrons and protons first into deuterium and then into helium.

What if 10^{10}-gm black holes were present at this time? These holes evaporate completely in roughly 3 minutes, which is why we consider them here. They have a Hawking temperature of 10^{16} degrees and would emit particles into the primordial background. (Admittedly, no one is sure what type of particle they would evaporate at such temperatures, but this is not so important.) These hot particles would strike and fragment the helium as it was being formed, turning it back into free neutrons and protons, which in turn would burn back into deuterium. For various technical reasons that concern the reaction rates, the nucleosynthesis chain now stops, no new helium is created, and the net result of injecting the black holes into our considerations is to increase the primordial deuterium abundance.

Now, astronomers sometimes think they know how much deuterium is in the universe. If it is all primordial, then arguments similar to those employed above for the 10^{15}-gm black holes show that 10^{10}-gm black holes could only have made up less than one-millionth the matter content of the universe at the time of primordial element synthesis. If black holes made up more than this fraction, then the amount of deuterium would be greater than the fraction observed by satellites. This calculation was first done by Zel'dovich *et al.* in the Soviet Union, and then by two of us (T. R. and R. M.).

Although primordial black holes of the kind we have described may or may not exist, the study of the quantum properties of black holes has a more general importance. In using quantum mechanics to show that any black hole has a finite temperature associated with it, Hawking took the first steps toward the unification of general relativity, quantum mechanics and thermodynamics. In the past, many of the greatest advances of physics have occurred when two or more previously distinct areas were united, as when Maxwell unified the electric and magnetic fields. The resulting science of electrodynamics is responsible for much of what we take for granted today.

It is unclear whether the unification of general relativity and quantum mechanics will produce anything "practical," such as black hole bombs, but it may eventually lead us out of the greatest current dilemma of relativity: the singularity. A singularity is a point in space-time at which all measurable quantities—temperature, pressure, gravitational forces—become infinite. When this occurs in a physicist's equations, there is simply no way to proceed and we say the theory has broken down. In classical relativity, such a singularity resides at the center of every black hole and, if we project the big bang back to time zero, the same singularity rears its ugly head—the equations describing the universe blow up. In classical relativity the singularity seems to be unavoidable. Furthermore, once in a singularity, there is no way out. Thus, relativity as it now stands disallows the beginning of the universe.

A simple argument shows, nonetheless, that classical relativity should not be the whole story at such early times in the universe's history. Earlier, we briefly discussed virtual particles, those particles which pop out of the vacuum of space and disappear. This phenomenon is allowed by the famous Heisenberg uncertainty principle, which states that there is a limit to the accuracy to which any quantity can be measured. That is, two particles of mass m can be created out of "nothing" as long as they don't remain in existence long enough to allow accurate measurement of their masses (hence the term "virtual"). In mathematical terms, the Heisenberg principle requires the two virtual particles to annihilate each other within a time of approximately $t \sim \hbar/2mc^2$. Here, $\hbar$ is Planck's constant, the natural constant that governs the scale of all quantum phenomena. Now, let us consider another time scale—the amount of time light requires to travel across a black hole. This will be roughly the Schwarzschild radius divided by the speed of light, or $t_s \sim 2GM/c^3$. If we set $m = M$ and require that $t = t_s$, we find that these times will be approximately equal at $M = \sqrt{c^3\hbar/G} \sim 10^{-5}$ grams, and $t = \sqrt{G\hbar/c^5} \sim 10^{-43}$ seconds. These are the formulae for the famous Planck mass and Planck time. What are their physical significance? The Schwarzschild radius of a black hole is typically the length scale at which gravitational forces become important. Thus, the light-crossing time is the time scale on which gravitational forces are important. Similarly, the Heisenberg formula tells us the typical time scale in which quantum effects dominate. By equating the two times (and masses) we have calculated the time and mass at which quantum effects have become just as important as gravitational effects. For instance, Hawking black holes already showed considerable quantum effects, but a 10^{-5}-gm black hole would evaporate completely in 10^{-43} seconds. This shows that quantum effects have overtaken those of gravity. These results tell us in particular that it doesn't make any sense to speak of black holes, as we know them, with masses less than 10^{-5} grams, and it doesn't make any sense to talk about a classical universe at times earlier than 10^{-43} seconds.

It is for exactly these simple reasons that most relativists believe quantum mechanics is necessary to rescue general relativity from the singularity at $t = 0$, a time even less than .0000000000000000000000000000000000000000001 seconds. And while it may be a long way to Tipperary, it is even a longer way to a full theory of quantum gravity. At this point, 10^{-43} seconds after the big bang, we will ourselves vanish in a quantum fluctuation, leaving only a smile (and a short appendix), for things have clearly gone out of sight.

APPENDIX

Communication with Observers
Falling into Black Holes

by A. C. Ottewill & Tony Rothman

This appendix gives a derivation of the answer cited in Section III of "Grand Illusions": how long can Richard and Tony undergo two-way communication while Richard is falling into a black hole? Here, we calculate an analytic expression for the "surface of last influence" of a Schwarzschild black hole. The surface of last influence is defined as the last time at which a distant observer can send a signal to a radially infalling observer and receive a reply in finite time.

The paper is technical and assumes at least an undergraduate course in general relativity. The exact numerical result is admittedly of little significance since the approximate answer can be guessed; however, the result does not appear to be in the literature, and the method of calculation—which is of greater interest—we have not found in any textbooks. We therefore present the calculation as an illustration for the student. Nonscientific readers are advised to skip the paper completely.

It is well known that to an observer orbiting at a fixed distance around a black hole, an object falling radially into the hole will appear to take an infinite time to reach the event horizon at $r = 2M$ (see eq. 5). The view is sometimes expressed that this fact makes discussion of the appearance of black holes to external observers meaningless because the surface of a collapsing star—itself an infalling object—never actually reaches the event horizon. That is, to the external observer, the surface of a 10-solar-mass star, for example, will collapse to within millimeters of the event horizon in a time of order of the *free-fall-collapse time* ~ 50 seconds, but will remain marginally outside the horizon forever.

Of course, to an observer falling *with* the surface of the star, the event horizon forms on the same free-fall time scale, and from that point he falls to the singularity $r = 0$ in a *proper time* $\tau < (\pi M)$. Nonetheless, a distant observer does *not* see the horizon form and, according to the above argument, a black hole never truly exists.

There are many fallacies in this argument; for example, after a *finite* time as measured by the distant observer, he can no longer communicate with an object falling into the hole. To see this, consider a mirror dropped out of the ship at a proper time $\tau = 0$. After some later *ship time* τ_{crit}, any photon emitted by the ship will reach the mirror only after it has crossed the event horizon, and hence cannot be reflected back to the ship. Although the existence of this "surface of last influence" is known (see Reference 1 at end of Appendix), its meaning does not seem to be widely appreciated, and an explicit value for τ_{crit} as a function of the ship's position does not, to our knowledge, appear in the literature. We present below a brief derivation giving such an expression.

It is easiest for this problem to work in Schwarzschild coordinates r, t. Because several r and t will be required, the diagram Fig. A is useful. We initialize both proper time and *Schwarzschild time* to 0 at the

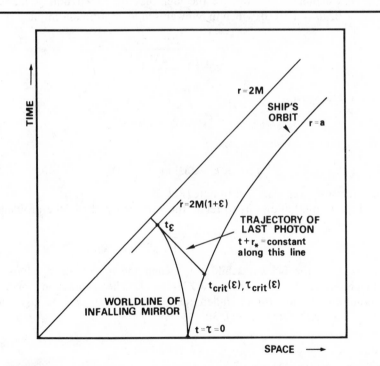

FIGURE A Schematic illustration of radial infall. A mirror is released at $t = \tau = 0$ and falls toward $r = 2M$. The time t_ε is the Schwarzschild time when the mirror reaches $r = 2M(1 + \varepsilon)$. The desired time t_{crit} for the ship is found by projecting back along the last photon trajectory. Along this line, the coordinate $t + r_* = $ constant.

moment the mirror is released. To find the desired ship-proper-time τ_{crit}, we will first find the corresponding Schwarzschild time t_{crit}. Then, for a stationary orbit, τ_{crit} is simply related to t_{crit} by

$$\tau_{crit} = \left(1 - \frac{2M}{a}\right)^{1/2} t_{crit} \tag{1}$$

where a will denote the radius of the ship's orbit.

Schwarzschild coordinates are singular at $r = 2M$. Hence, to find t_{crit}, we will consider a $t_{crit}(\varepsilon)$, which will be defined as the Schwarzschild time at which a photon is emitted *by the ship* such that this photon reaches the mirror when it has fallen to a distance $r = 2M(1 + \varepsilon)$. (See Fig. A.) Then $t_{crit} = \lim_{\varepsilon \to 0} t_{crit}(\varepsilon)$. With this procedure we can regulate any divergences in t_{crit}.

To calculate $t_{crit}(\varepsilon)$ we use the fact that on radial null geodesics $ds^2 = 0$, $d\theta = d\phi = 0$, which in Schwarzschild coordinates implies

$$dt^2 = \frac{dr^2}{\left(1 - \dfrac{2M}{r}\right)}.$$

Integrating gives

$$t + r_* = \text{constant} \tag{2}$$

for ingoing geodesics, where

$$r_*(r) \equiv r + 2M\ln(r/2m - 1) \tag{3}$$

is the usual "tortoise" coordinate (Ref. 2).

Equation (2) allows us to relate the various r and t of the problem. From Fig. A we have

$$t_{crit}(\varepsilon) + r_*(a) = t_\varepsilon + r_*(2M(1 + \varepsilon)). \tag{4}$$

Here, t_ε is the Schwarzschild time when the *mirror*, dropped from height a ($> 2M$), reaches $r = 2M(1 + \varepsilon)$. The expression for t_ε, assuming the mirror has fallen radially from rest, can be given in closed analytic form by (Ref. 3):

$$t_\varepsilon = 2M\ln\left|\frac{\left(\dfrac{a}{2M} - 1\right) + \tan(\eta/2)}{\left(\dfrac{a}{2M} - 1\right) - \tan(\eta/2)}\right|$$

$$+ 2M\left(\frac{a}{2M} - 1\right)^{1/2}\left(\eta + \frac{a}{4M}(\eta + \sin\eta)\right) \tag{5}$$

where $\eta = \cos^{-1}(2r/a - 1)$ is the usual cycloidal parameter, chosen so $t = 0$ at $\eta = 0$, i.e., at $r = a$. Note that as $r \to 2M$, $t \to \infty$, which is just the statement that the mirror requires an infinite Schwarzschild time to reach the event horizon, as mentioned in the introduction.

We need to evaluate t_ε at $r = 2M(1 + \varepsilon)$. Using

$$\tan(\eta/2) = \sqrt{\frac{1 - \cos \eta}{1 + \cos \eta}}$$

and working to first order in ε gives

$$\tan(\eta/2) \cong \alpha\left(1 - \frac{\varepsilon}{2\alpha^2}(1 + \alpha^2)\right) + O(\varepsilon^2)$$

where we have set $\alpha = (a/2M - 1)^{1/2}$.

Because the second term in (5) is finite as $\varepsilon \to 0$, we have

$$
\begin{aligned}
t_\varepsilon = 2M\ln &\left| \frac{\alpha + \alpha\left(1 + \frac{1}{2}\left(1 + \frac{1}{\alpha^2}\right)\right)\varepsilon}{\alpha - \alpha\left(1 + \frac{1}{2}\left(1 + \frac{1}{\alpha^2}\right)\right)\varepsilon} \right| \\
&+ 2M\alpha\left\{\cos^{-1}\left(\frac{4M}{a} - 1\right) + \frac{a}{4M}\left[\cos^{-1}\left(\frac{4M}{a} - 1\right)\right.\right. \\
&\left.\left. + \sin\cos^{-1}\left(\frac{4M}{a} - 1\right)\right]\right\} + O(\varepsilon).
\end{aligned}
$$

Simplifying gives

$$
\begin{aligned}
t_\varepsilon = -2M\ln\varepsilon &+ 2M\ln\left|\frac{4\alpha^2}{1 + \alpha^2}\right| \\
&+ 2M\alpha\left(1 + \frac{a}{4M}\right)\cos^{-1}\left(\frac{4M}{a} - 1\right) + 2M\alpha^2.
\end{aligned}
\tag{6}
$$

Using eq. (3) for $r_*(a)$ and $r_*(2M(1 + \varepsilon))$ yields

$$r_*(a) = a + 2M\ln(a/2M - 1)$$

and

$$r_*(2M(1 + \varepsilon)) = 2M(1 + \varepsilon + \ln\varepsilon). \tag{7}$$

It is now straightforward to find $t_{\text{crit}}(\varepsilon)$ by using (6) and (7) in (4). We note that the $\ln\varepsilon$ terms cancel, killing the divergence at $r = 2M$, as desired. Taking the limit $\varepsilon \to 0$ gives

$$t_{\text{crit}} = \lim_{\varepsilon \to 0} t_{\text{crit}}(\varepsilon) = 2M\left(\frac{a}{2M} - 1\right)^{1/2} \times \left(1 + \frac{a}{4M}\right)\cos^{-1}\left(\frac{4M}{a} - 1\right) + 2M\ln\left(\frac{8M}{a}\right).$$

(8)

Finally, we have

$$\tau_{\text{crit}} = \left(1 - \frac{2M}{a}\right)^{1/2} t_{\text{crit}}.$$

(9)

We recall that τ_{crit} is the proper time interval as measured on the ship between dropping the mirror and emitting the last photon that can be reflected back to the ship. Although the general form (8) is slightly clumsy, in the limit $a \gg 2M$, it reduces to

$$\tau_{\text{crit}} \cong t_{\text{crit}} \cong \pi M\left(\frac{a}{2M}\right)^{3/2},$$

which is $2^{1/2}/8$ times the period for circular Keplerian orbits

$$t_{\text{orbit}} = 2\pi\left(\frac{a^3}{M}\right)^{1/2}.$$

One intuitively expects $\tau_{\text{crit}} \sim t_{\text{orbit}}$ because $t_{\text{orbit}} \sim t_{\text{collapse}}$ is the only time scale in the problem.

In the limit $a/2M \to 1$, we get the peculiar result

$$t_{\text{crit}} = 2M\ln 4.$$

This is merely an artifact of the pathology of Schwarzschild coordinates at $r = 2M$. We note that the physically meaningful proper time τ_{crit} vanishes in this limit, as expected.

In conclusion, while the event horizon does take an infinite amount of Schwarzschild time to form, an orbiting observer can only hope to receive replies to messages sent before τ_{crit}, which for large orbits is $\tau_{\text{crit}} \sim 0.2 t_{\text{orbit}}$. The reply to any signal sent after this time will necessarily fall into the singularity at $r = 0$ and will never be received by the distant spaceship.

References

1. Charles Misner, Kip Thorne, John Wheeler, *Gravitation* (W. H. Freeman and Co., San Francisco 1973), pp. 873–879.
2. *Ibid.*, p. 663.
3. *Ibid.*, p. 666.

4

The New Neutrinos

by Richard Matzner & Tony Rothman

Once or twice in a decade a discovery is made which is so exciting that normal work abruptly halts, long-distance phone lines are set buzzing and corridors become alive with discussion. Then the race is on to get papers out. These are truly the best times to be a physicist. Such was the atmosphere in April 1980 when suddenly, from half a dozen research groups, came the startling declaration that the neutrino had a rest mass. Since then the excitement has died down somewhat; at the time of this writing, April 1985, we still do not know whether the neutrino has a mass. As a result, the following article has remained more or less current. I have added a postscript describing some new developments.

"The New Neutrinos" appeared previously in the May 25, 1981 issue of Analog.

In April of 1980, a wave of excitement swept through the physics community which, unlike many such tremors of the past, actually spilled onto the front pages of major newspapers and into the clutches of *Newsweek* and *Time* magazines.

The April event was, of course, the announcement that the elementary particle called the neutrino was not, as had been suspected for at least 30 years, massless. As it is wont to do, the popular press tried its best to transform a tentative result into reality: "New View of Universe." "Revolutionary changes in physics theory will be necessary." "Theological consequences."

In this article, we would like to take a more sober approach. We will explain what a neutrino is, why it was previously thought to be without mass and why some physicists now think otherwise. We will also discuss some implications that massive neutrinos have for physics. (We use the word "massive" in this article in the sense of "having mass" rather than "large.") You are warned that experiments involving neutrinos are notoriously difficult and that in another five years the next generation of experiments may refute the current findings, relegating us to obsolescence. We leave it as an exercise for you to discuss theological consequences.

The existence of the neutrino is intimately connected with the phenomenon of beta decay, and it is therefore impossible to understand one without the other. You may recall that radioactivity was discovered in 1896 by Henri Becquerel. He found that photographic film became exposed when placed in contact with various uranium compounds, even when both the film and the ore were housed in a completely dark drawer. Today we know that Becquerel's film was exposed by photons given off by the decaying uranium. Today, such high-energy photons are commonly produced by X-ray machines for medical purposes.

Over the next few years, work by Becquerel, Rutherford, the Curies and others showed that various substances, particularly radium, emitted three distinct types of radiation: alpha, beta and gamma. These designations, which survive to the present day, were at that time not so much scientific labels as measures of ignorance. Alpha radiation was stopped by a piece of paper, beta rays by a thin sheet of metal and gamma rays by a centimeter or so thickness of lead. At the time, nothing else was known about them and certainly the nature of their constituents was a mystery. Several decades of fumbling about were necessary to identify alpha rays as helium nuclei (two protons and two neutrons), beta rays as electrons and gamma rays as high-energy photons.

The beta radiation always posed a special problem. By the 1920s, the mass and charge of the beta particle were known to be the same as the mass and charge of an electron. Thus, physicists reasonably assumed that the two particles were identical; this later turned out to be correct. Nonetheless, given the identification of the beta particle with the electron, a serious problem yet remained for the physicist before 1930. Energy seemed to be vanishing entirely in a certain type of radioactive process called beta decay—the process in which decaying nuclei emit beta particles. To understand this problem, let us place ourselves in the position of an experimental physicist of that time.

We know that beta particles are electrons being emitted from decaying radioactive nuclei. Take a Geiger counter and place it near a radium sample. Each click of the Geiger counter represents one beta particle being given off by the radium. Furthermore, a Geiger counter connected to an "energy spectrometer" can be used to detect beta particles of many different energies. Thus, we can find out how many betas are given off at one energy, then a second energy, a third and so on. If we plot the number of emitted particles versus the energy at which they are emitted, we obtain a result like that shown in Fig. 1. This graph is called a beta decay spectrum and will be central to much of what we have to say. The x-axis simply gives the energy of the measurement and the y-axis tells us how many particles were emitted at that energy. You can see that at low energies a few particles are emitted; the number of particles climbs to some most probable energy, and then declines until, above a certain energy labeled E_{max}, no electrons are emitted at all. Being good 1920s physicists, we are very surprised at this result. Why?

In a uniform sample, all the radioactive atoms have the same energy to release in atomic disintegrations. In fact, the energy available for release is just E_{max} in Fig. 1. E_{max} is the energy of the most energetic electrons ejected from the sample; no electrons with higher energy are emitted, and thus it seems likely that those electrons are carrying off *all* the energy the atom has available for disintegration. Therefore, 1920s physicists at least tentatively identified E_{max} with the total energy available for release.

This brings us to the surprising point: If the nucleus decays, giving up E_{max} worth of energy and only emits one particle—the beta—then

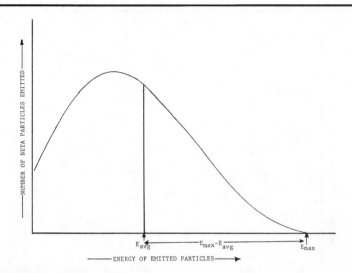

FIGURE 1 A typical beta decay spectrum. Electrons are emitted from the radioactive sample at all energies up to E_{max}. Those electrons emitted at E_{max} carry away all the energy available for decay. Ellis and Wooster trapped the escaping electrons in a calorimeter and found their average energy to be E_{avg}. Thus, the energy $E_{max} - E_{avg}$ seemed to be vanishing.

simple arguments based on conservation of energy and momentum show that all the betas should be emitted with precisely the same energy, E_{max}. After all, there is no place this energy can go except into the betas. Hence, the spectrum should not look like what we found in Fig. 1, but something like Fig. 2, a spike at E_{max}. This means that in the real life spectrum (Fig. 1) the electrons with energy less than E_{max} have carried away only a fraction of the available energy. There seems to be a substantial amount of energy simply disappearing during radioactive decay.

The first thought might be that we are simply not measuring all the electron energy with our Geiger counters, that we have overlooked some. However, an experiment published in 1927 by Ellis and Wooster makes us suspect this hypothesis. Ellis and Wooster placed a radioactive sample inside a calorimeter with very thick lead walls that would be sure to capture all the electrons. The idea is simple. Electrons smash into the walls of the calorimeter, transferring their energy in the process, and thus cause the temperature inside to increase slightly. The results of the experiment indicated that most of the elec-

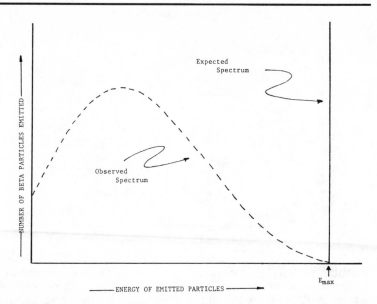

FIGURE 2 Conservation of energy and momentum led physicists before 1930 to expect all the electrons to be emitted at E_{max}, giving a spectrum like that shown here. Instead, they observed a spectrum as in Fig. 1, indicated here as a dashed line. This was one of the central puzzles in physics at that time.

trons captured in the calorimeter wall had an average energy much less than E_{max} (see Fig. 1). The difference between E_{max} and E_{avg} is the amount of energy that is vanishing into thin air.

You might get a feeling for the seriousness of the dilemma if you realize that Niels Bohr himself suggested abandoning the conservation of energy in beta decay! The situation was saved in 1930 when the eminent physicist Wolfgang Pauli wrote to his colleagues explaining he would not attend a conference at Tübingen because he planned to attend the annual ball at the Swiss Federal Institute of Technology. In this letter he also happened to mention that the above problems could be resolved if one assumed the decaying nucleus emitted a *second* particle along with the electron. Nowadays, this might seem like a simple idea, since physicists are familiar with hundreds of elementary particles. But one must remember that in 1930 only *two* elementary particles were known—the proton and the electron. Not surprisingly, many people refused to take Pauli's suggestion seriously. Among those who did believe him was Enrico Fermi, who in 1934 christened

the new particle " neutrino " for " little neutron." (By the date of chris-
tening, the neutron had already been discovered by Chadwick in
1932.)

Pauli's new particle solved the missing energy puzzle very neatly. If
the decaying nucleus emits *two* particles instead of one, the laws of
conservation of energy and momentum allow many ways to add the
energy of the electron and neutrino together to get the total energy
available, E_{max}. Thus, in the decay process, the electron gets whatever
it gets (call it E_e) and the neutrino takes up the slack ($E_{max} - E_e$). This,
then, explains the continuous nature of Fig. 1. The electrons may now
be emitted at any energy E_e up to E_{max}, with the neutrino carrying off
the rest.

At the same time, the existence of the neutrino explains why the
calorimeter experiment did not detect as much energy as expected—
the neutrinos simply carried off the excess. But notice, here is the first
indication that the neutrino must be a very weakly interacting particle. It was not stopped by the heavy lead walls of the calorimeter. If
Ellis and Wooster had trapped it in their apparatus, they would have
accounted for all the decay energy. Contrast the neutrino's behavior
with that of the alpha particle, which can be stopped by a piece of
paper.

By 1950 further experiments showed more directly that the neutrino
must be massless, or very nearly so. When a nucleus decays, it is not
only energy that must be conserved, but momentum as well. Experimenters measured the recoil (momentum) of the nucleus during beta
emission and the momentum of the electron itself. The two momenta
did not quite add up to zero as they should have. The missing piece
was given to the neutrino. The important point is that this missing
momentum was, *to within the experimental accuracy of about 10%,
given by the relativistic formula for the momentum of a massless particle*,
(momentum) = (energy/velocity of light) or, in symbols $p = E/c$.

The result that the neutrino mass is very small (or zero) means that
we were essentially correct in assuming that the maximum electron
energy, E_{max}, is the total energy available in the disintegration. In the
decay process, which also emits a neutrino, the electrons ejected at
E_{max} are usurping all the decay energy, leaving none for the neutrino.
In the language of relativity, this implies that the neutrino is massless.

We must pause here to explain a bit of confusing terminology that
often leaves the layman bewildered. Relativity tells us the energy
inherent in a body, say a rock, is composed of two parts. The first is
called the rest mass or the rest energy, two terms denoting the same
thing (remember $E = mc^2$). The rest energy is just the energy that
would be released if the entire rock were converted into energy while

it sat at rest on a table. When the rock is moving, the energy of motion, or kinetic energy, is added to the rest energy. Hence, $E_{total} = E_{rest} + E_{kinetic}$. When a physicist says a particle is massless, he means the rest energy—or, equivalently, the rest mass—is zero. This implies that the particle is never at rest, which means, according to relativity, it must always be traveling at the speed of light.

Now, you might object that since $E = mc^2$, even if a particle has no rest mass and only kinetic energy, it should then have a mass given by $m = E_{kinetic}/c^2$. This is true; all moving particles have, in some sense, an "effective" mass associated with them. The problem lies, however, not in the physics but in the terminology. When a physicist says "massless particle," he means a particle with zero rest mass, always traveling at the speed of light.

Leaving the world of semantics for that of physics, we return to the 1950s. Although by this time experimentalists had established that neutrinos had very small or zero rest mass, no one had ever actually detected a neutrino. It is not hard to see why. Using a few basic equations, it is easy to show that before being stopped by a collision with an atom a typical neutrino would travel through roughly 4 light-years of lead! Is it any wonder Ellis and Wooster missed them in a calorimeter? One might think it hopeless to detect neutrinos in a mere earthbound laboratory. Occasionally, however, physicists are clever. If massive numbers of neutrinos could be produced, then the odds of detecting a single neutrino would be increased. This was the basic idea behind the classic experiment of Clyde Cowan and Frederick Reines in the 1950s. In the experiment, which took the better part of the decade, they set up detectors near a nuclear reactor and observed certain nuclear reactions inside the detectors that required that neutrinos be produced by the nuclear reactor. Their first results appeared in 1953, with refinements in 1956 and 1959.

So finally, by the late 1950s, neutrinos were an established fact: particles that were emitted along with electrons in beta decay and that, to experimental accuracy, always traveled with the speed of light and had zero rest mass.

The flurry of excitement in April of 1980 would have the public—and indeed, physicists in other specialties—believe that since the 1950s everyone believed neutrinos to be massless and then suddenly changed their minds. After all, within the space of a week research groups from France, Switzerland, the Soviet Union and the United States all announced the advent of massive neutrinos. This cascade of announcements is a textbook case of result amplification by stimulated emission of publicity. After the initial announcement by Reines,

Pasierb and Sobel that massive neutrinos had been discovered, the other groups were stimulated to announce theirs for fear of being left out in the cold.

Several points should be made. First, nothing in the theory itself specifies the neutrino mass; it could be zero or it could be the mass of a baseball. However, by the beta decay experiment, the mass must be less than E_{max}, presumably much less. On the other hand, the early work described above only showed the neutrino to be massless within the 10% of E_{max} accuracy of the experiments. This allows for a small but finite neutrino mass. Second, it is not correct to say that for 30 years physicists have assumed the neutrino mass to be zero. History is always more complicated than that. For instance, in 1952 Langar and Moffat published a paper that put an upper limit on the neutrino mass. Specifically, their experiment showed the neutrino mass could not be greater than 0.05% of the electron. But note the date: 1952 was before the neutrino was detected! Physicists were not taking zero mass neutrinos for granted.

By 1969 Karl Bergkvist had improved Langar and Moffat's results by a factor of 4. It should be understood that neither of these results *implied* a mass for neutrinos; they showed that the mass could not be above a certain value. (We will go into a few details of these experiments below.)

This, to a large extent, explains the excitement of April. For the first time experimenters claimed that their results implied the necessity of a neutrino mass, and that they knew what this mass was.

Most public attention has been focused on the "neutrino oscillation" experiment of Reines, Pasierb and Sobel. (Reines, it will be recalled, was the co-discoverer of the neutrino in the 1950s.) Neutrino oscillations were considered as long ago as 1957 by Bruno Pontecorvo, who later extensively developed the theory, and are thus often referred to as "Pontecorvo oscillations." By any name, a full understanding of the phenomenon requires very advanced quantum physics. We can, however, give a close analogy to make the oscillations seem plausible.

Consider two closely tuned piano strings or organ pipes that are sounded together. Each is emitting a wave with a frequency that is slightly different from that of the other. Our ear perceives a very slow rise in volume, followed by a decrease, followed by another rise—the entire cycle taking a second or so. This common phenomenon is known as a beat frequency (see Fig. 3). The beating occurs at the frequency that is the *difference* between the frequency of the two piano strings or organ pipes. If the two pipes are exactly in tune, the frequencies are the same and the beating disappears.

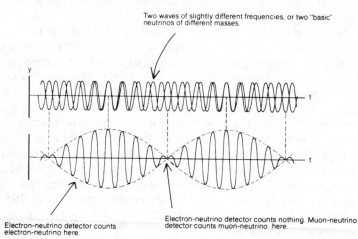

Two waves of slightly different frequencies, or two "basic" neutrinos of different masses.

y

t

t

Electron-neutrino detector counts electron-neutrino here.

Electron-neutrino detector counts nothing. Muon-neutrino detector counts muon-neutrino here.

FIGURE 3 The phenomenon of beats. Two sound waves of slightly different frequencies combine to produce a third wave with a beat rate given by the difference of the two original frequencies. A similar phenomenon occurs with neutrinos. Two basic neutrinos beat together to form an electron neutrino. Schematically, an electron-neutrino detector would detect the electron neutrino when the amplitude is large. When the amplitude is small the electron-neutrino detector would not record any neutrinos, but a muon-neutrino detector would.

A very similar thing happens with neutrinos. To see this, we must first mention that there are actually several different types of neutrinos. For the purposes of this discussion we first consider only two types: the electron-neutrino and the muon-neutrino. Now, it is known from quantum mechanics that particles behave like waves; a wavelength and frequency are associated with each particle. Thus, we can consider the electron- and muon-neutrinos to be waves.

Now, pretend there are two *other* types of neutrinos, more basic than the electron- and muon-neutrinos. These two basic neutrinos will behave just like the waves emitted from the organ pipes. If the frequencies of these two waves differ slightly, they will interfere and cause a beat frequency as before. We interpret this beating as the response of a detector designed to detect electron-neutrinos. When the "volume" is high, the two basic neutrinos have combined to form an electron-neutrino, which is detected. When the volume has gone down, the electron-neutrino has disappeared and a muon-neutrino has replaced it. At that point, a muon-neutrino detector would give a large response, and the electron-neutrino detector none. Thus, electron-neutrinos are changing into muon-neutrinos and back again. This phenomenon is the so-called neutrino oscillation.

The above analogy is actually quite close to the real story. We assume that the ordinary electron- and muon-neutrinos are composed of two more basic, unobservable neutrinos that are beating together. Now, we know from quantum mechanics that the frequency of a particle's wave is a measure of the particle's mass. (This isn't quite correct but will do for the purposes of our discussion.) Thus, the beat frequency, which is the difference in frequency between the two basic neutrinos, is really the difference between their masses. If the masses of the two basic neutrinos were zero, their frequency difference would be zero and there would be no oscillations. Consequently, if neutrino oscillations *do* exist, neutrinos are necessarily massive. (Note, however, that the converse is not true: lack of oscillations does *not* imply massless neutrinos. The neutrinos could have the *same* nonzero mass and the beat frequency would still be zero.)

Reines *et al.* claimed to have detected such oscillations as follows. They placed a detector near a nuclear reactor that was emitting electron-neutrinos (technically, electron *anti*neutrinos). In the detector two reactions could take place, the first of which could use either the electron- or muon-neutrino, the second of which required an electron-neutrino. If the electron-neutrinos were changing into muon-neutrinos, the rate of the second reaction would go down relative to the rate of the first reaction. This behavior is, in fact, what the Reines group has reported observing.

Unfortunately, although a CERN group performing a similar experiment feels that neutrino oscillations may exist, a Grenoble-Munich-Cal Tech collaboration has failed to confirm Reines' results. In addition, analysis of the data by Feynman and Vogel has resulted in a controversy over whether the data is as good as originally claimed. Furthermore, Feynman and Vogel point out that the current experiment was carried out at only one distance from the reactor. If electron- and muon-neutrinos are transforming into one another and back again, a process that occurs over a distance of several meters, one would expect to see different counting rates in the reactions if the detector were moved away from the reactor. Although this was done in an earlier experiment, with inconclusive results, it was not done in the more recent investigation. (Plans are underway to mount the detector on a railway car for just this purpose.) In light of these criticisms, Reines has evidently revised his original paper and resubmitted it to *The Physical Review Letters*. At the time of this writing, no one is in a position to say for certain whether neutrino oscillations really exist.

Recall that the beat frequency depends on the differences in mass of the basic neutrinos. Thus, although the existence of oscillations requires the neutrinos to have mass, the frequency of oscillations does not specify the individual masses, but only a mass difference. To determine a number for an individual neutrino mass, one needs a different type of experiment. It is the experiment performed by the Soviet team of Lyubimov, Nozik, Tretyakov and Kosik which has actually claimed to put a value on the neutrino mass. Interestingly enough, their experiment is just a refined version of that of Langar and Moffat and Bergkvist, which are all nothing more than a highly accurate measurement of the beta decay spectrum—in particular, the beta decay spectrum of tritium.

We are already well acquainted with this spectrum from Fig. 1, which was drawn for massless neutrinos. If we now assume that the neutrino has a small mass and calculate what the spectrum should look like, we arrive at Fig. 4. We see that the two spectra look almost identical except for a departure at the high-energy end. The departure is not difficult to understand. More energy is required to emit a massive neutrino than a massless one. This energy must come from somewhere and so is taken away from the electron. Recall that previously an electron emitted with energy E_{max} usurped all the decay

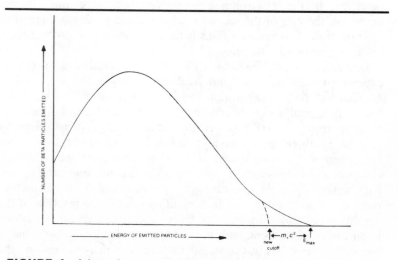

FIGURE 4 A beta decay spectrum showing the effect of massive neutrinos at the high energy end. The cutoff point has moved down from E_{max} to $E_{max} - m_\nu c^2$. The shape has also distorted slightly. The effect as drawn here is greatly exaggerated.

energy available and left none for the massless neutrino. If, however, neutrinos are massive, the neutrino must then be ejected with an energy corresponding to at least its rest mass, m_v. This means the electron, which formerly received E_{max} worth of energy, now only gets $E_{max} - m_v c^2$ worth. The upper end of the spectrum has thus been shifted by an amount corresponding to $m_v c^2$, the rest energy of the neutrino. This shift is indicated in Fig. 4. (You might wonder why the entire spectrum has not been simply shifted by the same amount. To answer this, one must look at the equation which quantum mechanics says describes the spectrum. This equation also shows, in addition to the shift already mentioned, a distortion of the spectrum for energies near E_{max}.) The job for the experimenter, then, is simply to determine whether the spectrum looks like Fig. 1 or Fig. 4.

We should not, however, leave you with the impression that these are easy experiments. In order to get the kind of accuracy required, extraordinary refinements are needed in the apparatus. For example, the tritium source must be no more than one molecule thick. Otherwise, electrons emitted from the bottom of the source would lose energy just by traveling through the tritium sample. This energy loss would be mistakenly interpreted as the mass of the neutrino. In addition, sophisticated focusing techniques are employed to make sure *all* the electrons emitted from the extended source will reach the Geiger counter. Otherwise, too few electrons will be counted to gather sufficient data. (Anyone who knows a bit about the theory of mass spectrometers should ponder the difficulties in this.)

The Soviets, who have been making these measurements for the past five years, claim their observations put the mass of the neutrino between 14 and 46 electron volts. Here, we are using the standard physicist way of talking about the mass of elementary particles—in electron volts, a unit of energy. One electron volt corresponds to a mass of about 10^{-33} grams. Hence, the mass of the proton, which is about 10^{-24} grams, is about a billion electron volts. The mass of an electron is about 500,000 electron volts.

We see that even if neutrinos do have a mass, it is fantastically small, roughly one ten-thousandth the mass of an electron or one hundred-millionth of a proton. If a new type of penny were minted from neutrinos instead of copper (which is made out of protons), a face falue of $1,000,000 of them would be needed before their weight equaled that of a standard Lincoln head. Here then is another indication of the sensitivity of the tritium decay experiments.

We might ask how such an incredibly small mass could possibly affect anything. At first glance there does not seem to be much difference between 14 electron volts and zero electron volts. Yet, it turns out, as always, that big surprises come in small packages. The rest of this article will be devoted to exploring some consequences of massive neutrinos.

One immediate result of the neutrino oscillations themselves is a partial explanation of the famous "solar neutrino paradox." The puzzle is simply put: conventional theory predicts that the sun should emit a certain number of neutrinos, while experiments by Davis *et al.* on earth observe one-fifth to one-third the predicted number. What is so paradoxical is that the experiment seems sound and the theory seems to be correctly applied. We should mention, however, that some physicists feel that the experiment is not quite as accurate as claimed, while others feel that the theory is incorrect. If either is the case, the "paradox" dissolves. If the paradox is genuine, an explanation must be found outside conventional solar physics. Many colorful hypotheses have been advanced: the sun does not burn by nuclear reactions; neutrinos are unstable and decay enroute to earth; a black hole resides in the sun's center and captures neutrinos.

Neutrino oscillations are a natural solution to the paradox which allows us to keep both the sun as we know it and the experiment. It turns out that the detector used by Davis *et al.* in their heroic experiment can only detect electron-neutrinos. But if electron-neutrinos are spending half their time as muon-neutrinos, then the detector is only half as likely to catch them. (Notice that this reasoning is similar to that employed in the Reines experiment.) In consequence, the recorded neutrino flux is only one-half the true value. This correction brings the experiment closer to the theoretical value but does not fully close the discrepancy—especially if there really is a factor-of-five difference. So it seems after all that neutrino oscillations may be only a partial solution to the paradox and another mechanism will have to be found for a full resolution. Perhaps the answer is that there are oscillations among more than two types of neutrinos.

Apart from their influence on solar physics, it is doubtful that massive neutrinos would play much of a role in any process that took place on a scale smaller than that of entire clusters of galaxies, tens to hundreds of millions of light-years across. This is because the neutrinos would be moving with a velocity too high to be captured by the gravitational field of a mere galaxy and would thus pass by this galaxy, hardly realizing it was there. Yet, the larger gravitational fields of clusters of galaxies would tend to make the neutrinos clump in their vicinity, and effects would be noticed.

For a long time it has been known that there is a "missing mass" problem with regard to clusters of galaxies. The orbital speed of individual galaxies in clusters is apparently too high to be due to the gravitational attraction caused simply by the other galaxies in the clusters. To produce the observed velocities, 3 to 10 times the observed mass is evidently needed. That is, if the observations of the galactic velocities are correct, most of the mass in a galactic cluster is invisible. Since no one has yet directly observed it—in the ways now possible for observing a galaxy—the name "missing mass" has been applied to this invisible source of gravitational attraction.

Now, if the neutrino mass is between 3 and 50 electron volts, then the random velocities of neutrinos would be slow enough to allow them to fall into the cluster under the cluster's gravitational pull. As David Schramm and Gary Steigman, among others, point out, these neutrinos offer a very plausible explanation of the missing mass—at least, it is plausible if you think massive neutrinos are plausible.

Researchers at the University of Texas Center for Relativity (Charles Evans, Nigel Sharp and we two) have noticed a further effect that massive neutrinos would have on the evolution of a galactic cluster. The process is simple and was discovered in 1943 by the distinguished astrophysicist S. Chandrasekhar. It is called dynamical friction and is closely analogous to the effect of air resistance on an orbiting satellite, say, the ill-fated Skylab. The space station is much more massive than the individual air molecules it collides with, but we all know the final result of such collisions: there are so many air molecules and the collisions occur so relentlessly that the spacecraft— alas—loses energy, starts to spiral toward the earth's surface and becomes the victim of *Time* magazine. As it spirals in, Kepler's laws say Skylab will go faster, since the orbital velocity is higher for a smaller orbital radius. Furthermore, the air below is denser than that above, so the collision rate goes up. Thus, the process goes quickly once it gets underway.

In Chandrasekhar's model of dynamical friction, it is not Skylab but an entire galaxy under consideration. Chandrasekhar investigated the effect of drag exerted on the galaxy in question by other galaxies in the cluster. The situation regarding galaxies is slightly different than that of earthly satellites; direct collisions aren't necessary since the long-range gravitational force of the galaxy itself drags smaller particles along. Suppose these smaller particles are massive neutrinos. If the total mass of the neutrinos in galactic clusters is great enough, we expect to see very strong evidence of dynamical friction working on the galaxies. Theoretically, the result could be a massive clump in the center of the cluster where all the galaxies—but for a few stragglers— have spiraled in.

Well, real clusters of galaxies don't look like this at all. They are more or less uniform in terms of the distribution of the galaxies contained in them. Hence, they cannot contain *too* much mass in the form of invisible small objects—be they black holes, massive neutrinos or *Time* magazines. How much mass is allowed before the effects of dynamical friction become evident? Roughly enough to explain the missing mass for the clusters that we discussed before. Expressed in terms of neutrino mass, the upper limit is about 10 electron volts per neutrino. This is consistent with both the Soviet experiments and the missing mass estimates, but since it is an upper limit only, it is also consistent with massless neutrinos.

We now turn to the most exciting implication of massive neutrinos—their possible effect on the evolution of the entire universe. There is at present a great debate over whether the universe is "open" or "closed." We will see that the question is not so simple as commonly supposed and that the addition of massive neutrinos does not make it any simpler.

The general theory of relativity allows for many possible universes. The so-called standard cosmology is the well-known big bang model: the universe began sometime in the past at arbitrarily high temperatures and densities. Since then it has been expanding and cooling down. Its expansion has also been slowing down because of the gravitational attraction of all its constituents—galaxies, dust, hydrogen and anything else. The question is whether the universe is slowing down enough to eventually recollapse in the "big crunch," or whether it is expanding with a velocity greater than the escape velocity of the system, in which case the universe will expand forever. The former case is called a "closed" universe, the latter an "open" universe.

The question of whether the real universe is open or closed can only be settled observationally. There are several avenues of attack, all of which have several levels of complexity. One can first measure the Hubble constant, a number that relates the distance to a galaxy to the velocity with which it is receding from us. This in itself is a difficult measurement, because distance measurements are difficult, and it is unclear whether the constant is known to be better than a factor of two. The Hubble constant is also a measure of the gravitational potential energy needed to stop the universe from expanding. Consequently, once the Hubble constant is known, we can calculate the amount of matter needed to close the observable universe. This theoretical value is given the label "critical density." The next step is experimental: we must decide whether the density of matter in the real universe is greater than the critical density. If so, the universe is closed.

This determination can be made by counting up all the material mass—the stuff in galaxies, typically—and deciding whether it exceeds this magic number. But this method always errs on the side of too little mass because there must be *some* invisible mass out there that was not counted. Nonetheless, most studies of this sort indicate that the density of the universe is roughly 20 times too little for closure, if one accepts the Hubble constant as being a certain value. On the other hand, P. J. E. Peebles has recently claimed that his studies of galaxies indicate there may be almost enough mass contained in them to do the trick.

A second method is to measure the so-called "deceleration parameter." Because of the finite time needed for light to traverse the universe, the light we see from distant galaxies was emitted earlier in the history of the universe's expansion than light emitted from nearby galaxies. We can measure the velocities of recession of each of these galaxies by observing its redshift. By comparing velocities of distant galaxies to those of nearby ones, we can determine the amount by which galaxies have been slowing down over the last few billion years. If the amount of deceleration is greater than a certain critical number, the universe is closed. This measurement is fraught with difficulties. In measuring the deceleration parameter, the distance to each galaxy must be calculated, which in turn is derived from a measurement of its apparent magnitude (brightness). The magnitude must then be corrected for the evolutionary process of the galaxy itself so that this evolution will not be confused with the slowing down of the universe as a whole. In short, little is known about such evolutionary processes and, at the present time, measurements of the deceleration parameter cannot be used to say anything about the closure of the universe.

There is yet a third method by which to decide the question. This method hinges on the ability of theory to calculate the nuclear reactions that presumably occurred shortly after the big bang, when the universe was about 3 minutes old and its temperature was 1 billion degrees Kelvin. If we assume that the elements currently observed in space were all formed in the big bang—with later contributions from nuclear processes in stars serving as a minor correction—then we can compare theory with experiment to deduce some of the properties of the early universe. We find, most importantly, that in order for the theory to produce the observed amounts of cosmic helium and deuterium, we are forced to make two significant statements about the universe.

The first conclusion is that there must have been about 10^9 photons for every proton at the time of element synthesis. This conclusion is important because it implies that the number of neutrinos is also about 10^9 per proton. Because the number of neutrinos has not changed since that time, nor has the number of protons, there are at present still 10^9 neutrinos per proton, or about 200 neutrinos for every cubic centimeter of space. The second conclusion is that the density of ordinary matter is now much less than needed to close the universe, by a factor between 15 and 50, depending on what value of the Hubble constant you choose to believe in.

The previous arguments, which are already complicated enough, were made under the assumption of massless neutrinos. It is on this littered stage that we now consider the opposite case: neutrinos with rest mass. As a recent award-winning essay by Schramm and Steigman emphasizes, the open universe conclusion drawn from the helium–deuterium argument cannot now be taken at face value. What changes need to be made? The statement that there must be 10^9 neutrinos for every proton remains valid. However, the previous estimates we have mentioned indicate that the probable neutrino mass is roughly 10 electron volts, or 10^{-8} times the proton mass. This, then, implies that the universe contains 10 times as much mass concentrated in neutrinos as in protons. This may be just enough to bring the matter up to the critical density and thus close the universe.

We have belabored the debate to show that the question of openness or closedness of the universe is not a simple one. As we have seen, the answer depends on the amount of matter in galaxies, on the cosmic isotope abundances and on the value of the Hubble constant. The universe may be closed even without the addition of massive neutrinos, although most studies indicate the contrary. Indeed some studies indicate the universe is so open that even massive neutrinos may not close it. The question is obviously difficult and may not be resolved for some time. In any case, those who thought the debate was closed are now forced to reopen it.

POSTSCRIPT

The previous article was written within a few months of the "Great Neutrino Scare of 1980." Surprisingly, the exposition still seems free of serious error. It is, however, incomplete. At the present time, Reines' experiment has not been successfully reproduced. Furthermore, on April 29, 1983 the Soviet team headed by Lyubimov removed their lower limit on the neutrino mass (that is, their results now allowed a zero neutrino mass). However, within six weeks the grapevine reported that they had reinstated their lower limit of roughly 20 electron volts. As we stated in the introduction, experiments involving neutrinos are notoriously difficult and the results of these experiments are still in flux. In addition, since the writing of this article many more roles for massive neutrinos have been proposed by theorists. Here I mention only a few.

We said in "On Cosmology" that the standard model of the universe has difficulty explaining the presence of galaxies. In general, the theory of galaxy formation is still rife with difficulties. One of the most serious difficulties is to get the matter in the universe to clump into the correct sizes within the current age of the universe. If neutrinos have a mass in the range we have been discussing, about 10 electron volts, they will begin to clump somewhat before the heavier particles that constitute the pregalactic gas. Hence, massive neutrinos may provide nucleation centers for the gas and enhance the rate of galaxy formation.

In another matter, it is known that the gas lying between galaxies—called the IGM, standing for Intergalactic Medium—is totally ionized. In other words, the electrons have been separated from the protons. According to conventional wisdom, quasar light causes the ionization. Recently, however, Patrick Osmer has given evidence that fewer quasars existed in the early universe than previously believed. If there existed too few quasars to ionize the IGM, then another mechanism will have to be found. Now, if neutrinos are massive, they will slowly decay into lighter particles and emit photons. Dennis Sciama has proposed that perhaps it is these photons from decaying neutrinos that ionize the IGM, not quasar light. Sciama has some impressive numbers to back his hypothesis, but in order for it to be conclusively established, it must be shown that quasars were indeed too few to ionize the IGM. This we have not been able to do.

Finally, the new Grand Unified Theories, which attempt to unify the electromagnetic and nuclear forces, predict a plethora of neutrino-like particles, all of them massive. The last few years have witnessed the advent of photinos and gravitinos, not to mention Goldstinos and Higgsinos. If it turns out that neutrinos do not have the mass required to accomplish the feats described in this article, it may be that massive photinos will do the trick equally well.

T. R.

5

Coincidences in Nature and the Hunt for the Anthropic Principle

by Bernard Carr & Tony Rothman

Recently, an idea that has become known as the anthropic principle has begun to capture the public's imagination. The idea itself is at least 20 years old but only recently have physicists taken up its exposition. Books by Paul Davies, Reinhard Breuer and others have by now appeared on the subject in Europe, and our colleagues Frank Tipler and John Barrow are co-authoring a text soon to be published by Oxford University Press. The difficulty in popularizing the anthropic principle is that it inherently deals with numbers—and the layman does not like numbers. Nonetheless, when Bernard Carr was visiting Texas in 1979, I suggested that we make an attempt at a popular article. Bernard agreed. After several publication delays, a shorter version of this article appeared in the October 26, 1981 issue of Isaac Asimov's Science Fiction Magazine. *The delays notwithstanding, I believe it was the first popular article on the anthropic principle to appear in the U.S. Here we print the original long version of the article. It contains more numbers and is consequently more difficult, but it explains more, as well. It is based in part on an article by Carr and Martin Rees, "The Anthropic Principle and the Structures of the Physical World," that appeared in the April 12, 1979 issue of* Nature.

Bernard Carr received his PhD under Stephen Hawking at Cambridge University in England, where his work concerned primordial black holes. Since then he has investigated pregalactic stars and cold big bang models of the universe. For enjoyment he plays the piano, and he is an expert on parapsychology.

NOTE: *This article is about some important numbers and some curious relationships between them. While the numbers are exact, the relationships are not. Throughout the discussion we speak in "orders of magnitude." Factors of two and five are totally insignificant, and even factors of 10 are only marginally troublesome. If an answer is off 100 times or 1,000, then we start to pay closer attention. Therefore, the diligent reader may lay aside his pocket calculator and relax; nobody will be checking his figures to 17 decimal places.*

Man is the measure of all things. This well-known dictum was first pronounced over 2,000 years ago by the Greek sophist Protagoras. It is not surprising that a Greek should be responsible for the sentiment. We need only recall the figures of Zeus, Apollo and the rest of the Olympic pantheon to recognize the face of man in god. The Greeks clearly placed man atop a very high pedestal.

Consider another dictum: The universe exists independently of man's awareness of it. This statement is the fundamental tenet of mechanism. It is a far cry from the Greek view of the world. Man has been reduced to insignificance. He is the merest speck in a galaxy that itself is only a speck in a universe filled with billions of galaxies. The laws of this universe operate as they have operated for all times past and as they will for all times to come—whether man is witness to their functioning or not. Energy is conserved whether man exists or not; stars shine whether man exists or not; the universe is as it is whether man exists or not. We can go to extremes: *Any* large-scale feature of the universe would remain exactly the same as it is if man suddenly vanished altogether.

But let us think a moment about this vast, impersonal, mechanistic universe. Let us ask the child's question, "Why is the universe as big as it is?" Most scientists would respond with a mechanistic answer: "The universe began with a big bang and is expanding." At any given time, the size of the observable universe is the distance light has traveled since the big bang, or roughly the speed of light multiplied by the age of the universe. Since the universe's present age is, say, 12 billion years, the present radius is about 12 billion light-years. Inherent in this straightforward answer is the belief that there is no compelling reason that the universe has a radius of 12 billion light-years other than that the universe is 12 billion years old.

There is, however, another answer to this question, an answer Protagoras might have liked better. Robert Dicke of Princeton University first gave it in 1961. "The universe," he said, "must have aged sufficiently for there to exist elements other than hydrogen. It is well known that carbon is required to make physicists." The carbon of which Dicke speaks, as well as other similar elements, is produced by cooking inside main-sequence stars. This process takes several billion years. Only after this time can the star explode into a supernova, scattering the newly baked elements throughout space, where they may eventually become part of life-evolving planets. So we see that to produce life, the universe must be at least several billion years old. Furthermore, the universe cannot be much older than this, or else all the material would have been processed into stellar remnants and man would have vanished from the scene.

Why then is the universe as big as it is? Because if it were much smaller or larger, we wouldn't be here to observe it! This startling conclusion turns the mechanistic answer on its head. The very hugeness of the universe, which seems at first to point to man's insignificance, is actually determined by his existence! This is not to say that the universe *itself* could not exist with a different size; rather, that the only universe we can observe is a universe of the present size.

By this argument, at least one feature of the universe—its size—is very much dependent on the awareness of man. We seem to have rediscovered something of the Greek point of view. Perhaps man is indeed necessary to explain certain features of the universe. It is appropriate that this conjecture has become known as the "anthropic principle."

We will not be surprised if you are left unconvinced by the above reasoning. After all, it could be "just a coincidence" that the age of the universe happens to be about the time needed to produce intelligent life. Our aim is to convince you that, coincidence or not, many additional features of the universe can be accounted for by use of the anthropic principle. Indeed, the evidence for the anthropic principle lies entirely in the large number of "coincidences" in physics that seem to be prerequisites for life.

Coincidences, everyone will agree, are very fascinating things. You are thousands of miles away from home in a flea-bitten flophouse in Afghanistan and you meet someone who knows your sister! You pick up a phone to call a friend cross-country and he is already on the line, having called you at the same instant! On the other hand, coincidences are liable to wither away upon closer inspection. Perhaps your sister sent her friend to spy on you. Perhaps you and your friend were *scheduled* to call each other. Once a coincidence has been explained, of course, it ceases to be one, and a good scientist will always try to understand an apparent coincidence in terms of what he already knows.

In this article we will be discussing a number of coincidences that crop up in physics. Our job will be twofold. First we must separate the apparent coincidences from the real coincidences. As remarkable as these apparent coincidences may seem at first glance, they can be explained by known physics. No anthropic principle is needed for their understanding. Once this has been done, a number of real coincidences remain. Examining these coincidences, which seem to be necessary for the existence of life, will be the second part of our job. Because conventional physics has been unsuccessful in explaining them, we turn to a possible anthropic answer. Both the apparent and real coincidences of nature are of a very different type than the above examples. They do not involve flophouses or phone calls. They involve the constants of nature themselves and have set the physicist on the hunt for the anthropic principle.

Some Basic Numbers

Certain numbers appear and reappear in the most important formulas of physics. They appear so frequently that a good physicist can often guess the solution to a problem just by considering which of these numbers will play a crucial role. Just as important as their numerical values are their "dimensions," expressed in terms of mass, length and time. The dimension of a number tells us what sort of quantity the number measures. We now list the most important of the "natural constants" with their approximate values and their dimensions, which will be expressed in the cgs system of centimeters, grams and seconds. The table on page 112 gives a more complete listing of the constants and combinations of constants used in this article.

c = the velocity of light = 3×10^{10} cm/sec

$\hbar$ = Planck's constant = 10^{-27} gm cm^2/sec

G = the gravitational constant = 7×10^{-8} cm^3/(gm sec^2)

m_p = mass of the proton = 2×10^{-24} gm

m_e = mass of the electron = 9×10^{-28} gm

e = charge on the proton or electron = 5×10^{-10} (gm cm^3/sec^2)$^{1/2}$

Certain combinations of these constants have a special physical significance. For example,

$$r_p = \hbar/m_p c = 10^{-13} \text{ cm} \qquad (1)$$

is the so-called Compton wavelength of the proton or, roughly speaking, its size. If we divide the "size" of the proton by the velocity of light, we get a number

$$t_p = \hbar/m_p c^2 = 10^{-23} \text{ sec}, \qquad (2)$$

Table of Fundamental Constants and Other Important Numbers

SYMBOL	NAME	SIGNIFICANCE	VALUE IN CGS UNITS (APPROXIMATE)
c	speed of light	—	3×10^{10} cm/s
$\hbar$	Planck's constant	governs scale of all quantum phenomena	1×10^{-27} gm cm^2/sec
G	gravitational constant	governs strength of gravitational force	7×10^{-8} cm^3/gm/sec^2
m_p	mass of the proton	—	2×10^{-24} gm
m_e	mass of the electron	—	9×10^{-28} gm
e	charge of the electron or proton	—	5×10^{-10} (gm cm^3/sec^2)$^{1/2}$
$r_p = \hbar/m_p c$	Compton wavelength of proton	roughly the size of the proton	10^{-13} cm
$t_p = \hbar/m_p c^2$	proton timescale	time for light to travel across a proton	10^{-23} sec
$a = \hbar^2/m_e e^2$	Bohr radius	roughly the size of an atom	10^{-8} cm
$\alpha = e^2/\hbar c$	fine structure constant	governs strength of electric force between two protons	$1/137$
$\alpha_G = Gm_p^2/\hbar c$	gravitational fine structure constant	governs strength of gravitational force between two protons	5×10^{-39}
f^2	strong force fine structure constant	governs strength of strong nuclear force	15
α_w	weak force fine structure constant	governs strength of weak nuclear force	10^{-10}

which is the time required for light to travel across the proton. A slightly more complicated combination,

$$a_0 = \hbar^2/m_e e^2 = 10^{-8} \text{ cm},\tag{3}$$

is called the Bohr radius and specifies roughly the size of an atom. If we divide the proton mass (m_p) by the volume of an atom (a_0^3), we get a density (ρ_0). In terms of the fundamental constants, we can write this density as

$$\rho_0 = m_p/a_0^3 = m_p m_e^3 e^6/\hbar^6 = 1 \text{ gm/cm}^3.\tag{4}$$

This is the atomic density and, to within a factor of 10, it characterizes the density of all solid and liquid objects. The astute reader may be surprised that this very complicated combination reduces to the simple number 1. This is, in fact, a genuine coincidence; it cannot have any physical significance because if we used another system of units (such as the mks or the English system), ρ_0 would not be 1. We mention this to stress that *some* coincidences are just what their name implies.

Now, it is a favorite pastime of many physicists to engage in numerology. You might like to join in the fun by picking several of the numbers c, $\hbar$, G, e, m_p, m_e and forming as many "dimensionless" combinations from them as you can. By "dimensionless" we mean that all the units of the combination have canceled out and we are left with a "pure number." No centimeters, grams or anything else remain. An example of such a combination is the so-called fine structure constant:

$$\alpha = e^2/\hbar c = 1/137.\tag{5}$$

Only this number (or powers of it) can be formed from powers of e, $\hbar$ and c such that all the units cancel. You might guess that α is of particular significance since it is composed of three fundamental constants. Indeed, it plays a crucial role in any situation where the electromagnetic force is important. Whether because of this or more metaphysical reasons, the physicist Eddington once posited the existence of 137 degrees of freedom in the universe.

Another important dimensionless number is the *gravitational* fine structure constant:

$$\alpha_G = \frac{G m_p^2}{\hbar c} = 5 \times 10^{-39}.\tag{6}$$

The fact that α_G is so much less than α reflects the fact that the gravitational force between two protons is much smaller than the electric force between two protons. Gravity dominates the structure of large bodies like stars only because large bodies tend to be electrically neutral and the electric forces cancel out.

As an additional exercise, you might want to confirm that, using powers of $\hbar$, c and G, the only quantities that can be formed with the dimensions of mass, time and length are:

$$M_{\text{Pl}} = (G/\hbar c)^{-1/2} = 10^{-5} \text{ gm};$$

$$t_{\text{Pl}} = (G\hbar/c^5)^{1/2} = 10^{-43} \text{ sec};$$

$$R_{\text{Pl}} = (G\hbar/c^3)^{1/2} = 10^{-33} \text{ cm.} \qquad (7)$$

The subscript "Pl" denotes (Max) Planck, who is credited with the discovery of these combinations. The Planck scales, as they are called, have some very interesting properties that we will mention later. Note that they can themselves be expressed in simple powers of the gravitational fine structure constant multiplied by the mass and length scales of the proton:

$$M_{\text{Pl}} = \alpha_G^{-1/2} m_p; \qquad R_{\text{Pl}} = \alpha_G^{1/2} r_p; \qquad t_{\text{Pl}} = R_{\text{Pl}}/c. \qquad (8)$$

It may not seem very remarkable that the Planck scales can be expressed in terms of simple powers of the gravitational fine structure constant since they were all constructed out of the same fundamental constants to begin with. However, the remarkable thing is that the size of nearly every naturally occurring object can also be expressed in terms of α and α_G. Because of this, we can predict a large number of surprising relationships between the different scales of structure in the universe. These relationships are so surprising that at one time many were thought to be genuine coincidences.

We will now give a few examples to illustrate how these dependences of natural scales on α and α_G arise. You should bear in mind, however, that the following examples will be very simplistic; we will only use "order-of-magnitude" arguments that give answers to within factors of about 10. To determine the scales more precisely, one needs to use more detailed physics (and consequently more work) than the present discussions will require. Nonetheless, order-of-magnitude calculations are often surprisingly accurate. The kind of complex calculations that have made physics legendarily difficult usually only refine the results by a factor of two or three. The reader can rest assured that even if the physics here is not quite on the mark, it is at least reasonable. In this spirit, all remaining equations will involve the " $\sim$ " (usually called "twiddle") rather than the " $=$ " sign. This means that the quantities on either side of the twiddle are not precisely equal, but lie within an order of magnitude of one another. (We should properly have used the twiddle in a few previous expressions.) The art of determining things to an order of magnitude, "twiddleology," can give great insight into physics. It allows you to see the forest for the trees or, in our case, the galaxy for the stars.

Stars, White Dwarfs, Pulsars and Black Holes

We'll start off with a detailed discussion concerning the allowable mass of stars. A famous coincidence is that the mass of the sun is very nearly given by the following relation:

$$M_\odot \sim \alpha_G^{-3/2} m_p. \tag{9}$$

That the mass of the sun should depend on such a simple power of the gravitational fine structure constant is not obvious *a priori*. In fact, at least one elaborate cosmological theory has been invented to explain it. We now know that stars are expected to have a mass of this order, so it is no longer considered coincidental. What is the explanation?

A star is a cloud of gas, mostly hydrogen, that has collapsed under its own gravity. Not every collapsing cloud of gas will become a star. If the temperature of the gas rises enough during the collapse for nuclear reactions to commence, then the hydrogen in the gas cloud ignites and a star is born. At this point the heat released by the nuclear burning of the hydrogen causes an outward pressure that prevents the new star from collapsing any further. The star now begins its "main-sequence" lifetime, a stage that we will discuss shortly. What happens if the ignition temperature of hydrogen is never reached? In this case the cloud continues to collapse until further compression is halted by a peculiar quantum mechanical effect called "degeneracy pressure." If a cloud succeeds in collapsing this far, it just forms a planet. What we need to decide, then, is when a gas cloud will collapse to a star and when it will collapse to a planet. To find the dividing line, we must discuss the phenomenon of degeneracy pressure.

Degeneracy pressure is a direct result of the famous Pauli exclusion principle, which states that certain types of particles—such as electrons and neutrons—cannot be pushed too closely together without jiggling about. The closer they are pushed together, the more violently they jiggle. One might say the particles get very agitated when they lack elbow room and that this agitation causes an outward pressure. Now, energy is required to make the particles jiggle, and this energy comes from the gravitational potential energy released as the cloud collapses. While the particles remain far apart, most of this released energy goes into *heating* the gas cloud. However, once the particles get close enough for the exclusion principle to go to work, the energy is diverted into the jiggling, and the star becomes *cooler* as it collapses further. The maximum temperature attained depends on the mass of the cloud (M). It can be shown that this temperature will exceed the ignition temperature of hydrogen when M is about one-tenth of the mass specified by relation (9). In other words, any gas cloud of more than one-tenth the mass of (9) will collapse into a star. Any cloud smaller than this limit cannot ignite its nuclear fuel and so, as mentioned earlier, it will collapse until degeneracy pressure halts further compression at the planetary scale. We will defer discussion of planets until we have carried stars through to their death throes.

Once a star has ignited, it begins its main-sequence lifetime. The main sequence is an equilibrium state in which the outward pressure caused by burning balances gravity, and the rate at which the star cools is balanced by the rate at which heat is generated by nuclear reactions. The main-sequence stage lasts until the star has burned all the hydrogen in its core. Not all stars make it intact through the main-sequence state. For reasons too technical to go into here, stars larger than about 10 times $\alpha_G^{-3/2}m_p$ tend to be unstable to pulsations and disrupt long before they finish hydrogen burning. Thus, only gas clouds with a mass in the narrow range of 0.1–10 solar masses can form stable, main-sequence stars—that is, those clouds with masses roughly specified by (9). As an additional note, the main-sequence lifetime for large stars turns out to be given by a remarkably simple formula:

$$t_{ms} \sim \alpha_G^{-1}(M/M_\odot)^{-2}t_p. \tag{10}$$

We will use this result later. The important point is that we have explained (9).

What happens when the star has completed its main-sequence burning? At first, because there is no longer a pressure holding it up, the star will begin to collapse. As it collapses, the star will get hotter. It may even get so hot that the helium produced by the hydrogen burning will itself burn into heavier elements. Helium will burn first to carbon and then to successively heavier elements. When it has burnt all the way up to iron, the star can extract no more heat out of nuclear reactions. Once again the star begins to collapse from lack of supporting pressure. However, the collapse may still be stopped by the degeneracy effects discussed earlier. At first it will be the electrons that try to stop the collapse. They will succeed if the star's mass is less than a critical value called the "Chandrasekhar mass." (This limit is named after the man who worked it out in detail, although Landau arrived at the same basic conclusion several years earlier by an order-of-magnitude calculation. Such is the power of back-of-the-envelope physics!) The Chandrasekhar mass is of the same order as the mass that characterizes stars in general, that is, the mass given by (9). A more precise calculation gives the result $1.4M_\odot$. The collapse of a star whose mass is less than the Chandrasekhar limit will be halted at some point by electron degeneracy pressure. When this point is reached, the star becomes what is called a "white dwarf," with a radius of order 10,000 kilometers.

If the collapsing star is more massive than the Chandrasekhar mass, the degeneracy pressure of the electrons will not be sufficient to support the weight of the star. The collapse continues. Now the electrons get squashed into protons to make neutrons. Neutrons, like electrons, provide a degeneracy pressure, but a much stronger one. If the star is not too large, the neutron degeneracy pressure may now stop the collapse that the electrons failed to halt. In this case, the star ends its life as a "neutron star" with a nuclear density of roughly $\sim m_p/r_p^{3} \sim 10^{15}$ gm/cm^3. On earth we may detect such a star as a pulsar. The critical mass below which neutron stars can form is not as easy to calculate as the Chandrasekhar limit, but it is also about 1 solar mass.

It is time to pause briefly and take stock. Although we have not shown the calculations explicitly, we have indicated that main-sequence stars, white dwarfs and neutron stars all have masses of order $\alpha_G^{-3/2}$ times the proton mass. The reason that the same mass crops up in all three contexts is that all these objects have an equilibrium state determined by the balance between degeneracy pressure and gravity. The strength of gravity is measured by G, the gravitational constant. Quantum effects, like degeneracy pressure, always involve Planck's constant, $\hbar$. Thus, any equation purporting to determine the balance between gravity and degeneracy effects must involve the ratio $G/\hbar$, which measures, in some sense, their relative strengths. We are no longer so surprised at relation (9). The ratio $G/\hbar$ dutifully appears in it, hidden under the guise of α_G, the gravitational fine structure constant.

There exist stars too massive for even neutron degeneracy to support. These stars face one of two dramatic fates. For reasons still not well understood, such a star may explode as a supernova. In this case, all the heavy elements that were generated in the nuclear burning stage will be spewed out into space. The exploding star may be disrupted entirely or it may leave a surviving core that is small enough to avoid collapse: a neutron star. The alternative fate for the star is even more dramatic. It may continue to collapse indefinitely, until the star has been compressed to a point of essentially infinite density. All that is left behind as a vestige of the star that shone brightly for 10 billion years is a region of space where gravity is so strong that nothing—not even light—can escape: a black hole.

Black holes have been described in so many other popular articles (including some in this book) that we need not discuss them at length here. However, it is worth pointing out that the radius of a black hole of mass M (the Schwarzschild radius) given by the well-known formula

$$R = 2GM/c^2$$

can also be expressed as

$$R = 2\alpha_G(M/m_p)r_p. \tag{11}$$

The Schwarzschild radius associated with the sun can be written with the help of (9) as

$$R_\odot \sim \alpha_G^{-1/2} r_p. \tag{12}$$

This comes out to be about a kilometer. The fact that this is so small is a direct consequence of the smallness of α_G.

Black holes smaller than about $1 M_\odot$ would require greater compression for their formation than is likely to arise in the present epoch of the universe. They may, however, have been created in the first few moments of the big bang. If small black holes did form in the early universe, they would have very remarkable quantum properties. Stephen Hawking has shown that black holes emit particles, just as any object does that is at a finite temperature. (We usually think of hot bodies emitting radiation, but radiation can equally well be thought of as particles.) The temperature of a black hole, which is inversely proportional to the hole's mass, is only 10^{-7} degrees for a solar-mass hole, but 10^{11} degrees for a hole of 10^{15} grams. As a result of this particle emission, a black hole will completely evaporate in a time that can be expressed as

$$t_{evap} \sim \alpha_G^2 (M/m_p)^3 t_p. \tag{13}$$

So we see that yet another important process in nature is directly governed by the value of α_G. For $M \sim M_\odot$, t_{evap} is much longer than the age of the universe, but any hole smaller than 10^{15} grams would have evaporated by now. Note that a black hole of the Planck mass, defined in (7), has a Schwarzschild radius of the Planck length and evaporates in a time equal to the Planck time. There is also something special about the Schwarzschild radius of a 10^{15}-gram black hole. We will let you guess what it is until we return to that coincidence later.

Planets, Asteroids and Men

We have traced the evolution of massive gas clouds from their collapse into main-sequence stars to their deaths as white dwarfs, neutron stars or black holes. It is time to return to those smaller clouds with masses less than $0.1M_\odot$ which evolve into planets. If the cloud has a mass just under $0.1M_\odot$, it will evolve directly to the sort of white-dwarf state that eventually awaits its larger counterparts. If the cloud is *much* smaller than $0.1M_\odot$, it will evolve into a solid planet. The basic difference between a solid body and a gaseous body is that the structure of the solid is a result of the chemical bonds between its constituent atoms. These bonds have their origin in electrostatic forces whose strength is measured by α, the fine structure constant.

We have stated several times that the final equilibrium state of a star, be it a white dwarf or a neutron star, is determined by the balance between degeneracy pressure and gravitational attraction. In contrast, gravity does not play much of a role for small bodies like planets. Rather, the equilibrium state of a solid object is determined by the balance between degeneracy pressure and the electrostatic bonds. It is easy to show that this balance requires the density of the object to be on the order of the atomic density, given by (4). All solid objects—the earth, bricks or even animals—have roughly this density, although the precise value varies according to the sort of atoms the object is made of. Thus, gas clouds much smaller than $0.1M_\odot$ collapse until degeneracy pressure halts compression at the atomic density. Then they settle down as solid planets.

We indicated above that there is a mass range just below $0.1M_\odot$ in which planets form that are *not* solid, but similar to white dwarfs. It is in this mass range that gravitational effects become important. Thus, this size may be taken to be the maximum size allowable for a planet. A simple calculation shows this mass to be:

$$M_{max} \sim (\alpha/\alpha_G)^{3/2} m_p. \tag{14}$$

M_{max}, which is just $\alpha^{3/2}$ times the characteristic mass of a star—see (9)—is about the mass of Jupiter. Jupiter, therefore, lies on the dividing line that separates planets from stars. A larger planet turns into a white dwarf with more than atomic density. The fact that α appears in the formula reflects the crucial role played by the electrostatic force in determining the equilibrium state of a solid body.

To find the *minimum* size of a solid planet, we make the assumption that planets should be more or less round. This requires that they be larger than the size of their own mountains, otherwise we would have a very lopsided planet indeed, something we would call an asteroid. It is actually quite easy to estimate the maximum size of a mountain. We merely assume that a mountain cannot be so high that the pressure exerted by its weight on the planet's surface liquefies its base. That is, the pressure must not break the chemical bonds that make the base material solid. Simple calculations along these lines yield a maximum height for earth mountains of about 10 kilometers, in good agreement with what is observed. In general, we find that the maximum mountain size increases as the planet's mass decreases, so there is a critical mass *below* which the planet is smaller than its own mountains. This turns out to be:

$$M_{min} \sim A^{-2}(\alpha/\alpha_G)^{3/2}m_p \tag{15}$$

where A is the atomic number (the number of nucleons per atom) of the planetary material. We take this to be the maximum size of an asteroid. Anything larger is a planet. M_{min} is less than the M_{max} of (14) by a factor of A^{-2}, which is around 10^{-4}.

We can restrict the size of planets even further if we require that they be "habitable." Of course, it is very difficult to decide just what "habitable" means since we know of only one example of a habitable planet in the universe—the earth. We will make two rather general requirements. First, we require that the planet have an atmosphere with elements heavier than hydrogen. Second, we require a suitable temperature for biological reactions to take place. The optimum temperature, T, for organisms is related to the energy released in reactions between complex biological molecules. If T is too high, the molecules disrupt; if T is too low, the reactions take an enormously long time.

What are the results of these requirements? The fact that the planet is at a finite temperature causes the gases in the atmosphere to move with a "thermal velocity" characteristic of that temperature. The higher the temperature, the higher the thermal velocity. If the thermal velocity of a gas is greater than the escape velocity of the planet, the gas will escape from the planet's atmosphere. Now, at a given temperature, heavier molecules will have a lower thermal velocity than lighter ones, hydrogen being the lightest. We require that, at the biological temperature, the atmosphere must contain elements heavier than hydrogen. This means the hydrogen must be allowed to escape; in other words, its thermal velocity must be slightly greater than the escape velocity of the planet. (It must not be too much greater or *all* the gases would escape.) This gives a mass for habitable planets:

$$M_{life} \sim 10^{-3}(\alpha/\alpha_G)^{3/2}m_p \sim 10^{-3}M_{max}. \tag{16}$$

The number 10^{-3} derives from the "optimum" biological temperature. Only a small fraction of the total planets in the range between M_{min} and M_{max} would satisfy this condition.

William Press of Harvard has pushed this sort of argument further to estimate the maximum size of a planet-dwelling creature. Assuming that the planet has a mass, M_{life}, indicated above, Press argues that the creature must not be so large that it breaks apart when it falls. In other words, the gravitational potential energy released in its fall must not break the chemical bonds that hold the molecules of its body together. This yields an upper limit for a land-dwelling creature:

$$h \sim (\alpha/\alpha_G)^{1/4} a_0 \sim 10 \text{ cm} \qquad (17)$$

where a_0 is the Bohr radius. This expression is very roughly (order of magnitude, remember) the height of a man. Notice how simply this height depends on the ratio (α/α_G).

At this point, we have gone through many arguments and have many results. We repeat that, although the calculations behind the arguments are simple, they contain the essential physics and their results are reasonable. In order to collect these results and display them in a convenient form, we draw the accompanying figure. The diagram indicates the mass and length scales of the objects we have previously discussed (and a few more besides) not only in terms of

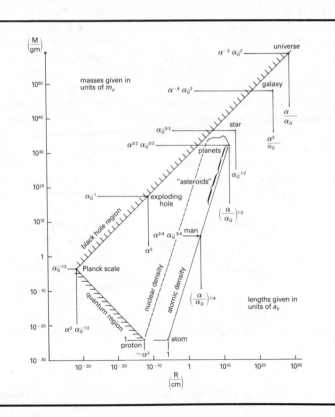

grams and centimeters, but also in units of the proton mass (m_p) and the Bohr radius (a_0). For example, we see the point marked "star" at about 10^{10} on the "length" axis. This means the radius of a star is about 10^{10} centimeters. But we also notice a vertical line indicating that this length is equivalent to the expression $\alpha_G^{-1/2}a_0$. Similarly, the mass of a star is indicated to be both 10^{33} grams and $\alpha_G^{-3/2}m_p$.

The diagram has some very interesting properties. Its most important message is that many, if not most, of the important length and mass scales of the universe can be expressed simply in terms of α and α_G. In consequence, we can discover a number of amusing relationships between the different scales. For example: the size of a man is the geometric mean of the size of a planet and the size of an atom; the size of a planet is the geometric mean of the size of an atom and the size of the universe; the mass of a man is the geometric mean of the mass of the planet and the mass of a proton. (The geometric mean of two numbers, x and y, is defined to be $(xy)^{1/2}$.)

It is worth stating once more that these relationships should *not* be regarded as coincidences, no matter how remarkable they may seem at first sight. Because we have explained the length and mass scales of these objects in terms of fundamental physics, these "coincidences" must be regarded as logical outcomes of the physical theory. In a way, it is not very surprising that the simple relationships indicated in the diagram arise. Any stable object in the universe reflects a balance between various forces. On scales larger than atoms, the only important forces are gravity, electromagnetism and quantum forces. Since α and α_G are the only dimensionless numbers that can be constructed from m_p, G, e and $\hbar$—along with the speed of light—it is not too surprising to find them surfacing in the sort of results presented here. There is, however, no *a priori* way of knowing which powers of α and α_G will appear in the various scales. These must simply be calculated.

You may justifiably be wondering what is so anthropic about all this. We have yet to use any property of man except his atmospheric requirements and his ability to break. Let us make the anthropic connection by returning to the question that began this journey: "Why is the universe as big as it is?" Instead of asking this precise question, we are going to ask a different version that, at first, might seem only distantly related. We are going to ask for the explanation of a very famous cosmological coincidence. It can be expressed as:

$$\frac{ct_0}{a_0} \sim \frac{\alpha}{\alpha_G} \sim 10^{37}, \tag{18}$$

where t_0 is the age of the universe. On the left of the first twiddle is the ratio of the size of the observable universe (ct_0) to the size of an atom (a_0); on the right is the ratio of the electromagnetic force between two protons (α) to the gravitational force between two protons (α_G). Although the size of the two numbers is too enormous to be readily conceived, they happen to be nearly the same. This is a real coincidence, one that cannot be explained by known physics.

We may feel compelled to explain this remarkable coincidence, or we may dismiss it as unimportant. The great physicist Paul Dirac felt compelled to explain it. He suggested that the gravitational constant, G (which appears in α_G), is not really a constant at all, but decreases as the universe expands in such a way that the coincidence remains true at all times. (You can easily work out the dependence of G on t in this hypothesis.) Dirac's theory may well be inconsistent with observational evidence, and if so, it will have to be thrown out.

There is another explanation for this famous coincidence. It is Dicke's original anthropic argument, given at the beginning of the article. Unlike Dirac, Dicke does not assume that the coincidence holds at all times. On the contrary, he claims, it only holds at one epoch—the present. But he further argues that life can exist *only* at the present. Therefore, any observer will *automatically* find coincidence (18) satisfied.

We can now check Dicke's argument in a more precise way. Recall that he reasoned that life could only exist when the age of the universe was comparable to the main-sequence lifetime of a star, in other words, when $t_0 \sim t_{ms}$. Is this observed to be true? We found an expression for t_{ms} in (10). It depends on α_G^{-1}. If we rewrite (18) as

$$t_0 \sim \frac{\alpha}{\alpha_G} \frac{a_0}{c} \tag{19}$$

we see that t_0 also depends primarily on α_G^{-1}, the other factors being less important. This is highly encouraging if we are to establish that $t_0 \sim t_{ms}$. However, we also see that t_{ms} depends critically on the mass of a star. We thus cannot claim $t_0 \sim t_{ms}$ in all cases. Nonetheless, for stars of about $1M_\odot$ or slightly less, t_0 and t_{ms} are comparable. In order to make Dicke's argument more convincing, we should have to establish that most stars are of this size. There do exist technical arguments indicating that this might, in fact, be true. In light of these arguments, the anthropic principle remains a tantalizing contender for the explanation of the size of the universe.

An interesting corollary coincidence results from the coincidence (18), which concerns evaporating black holes. Let us set the time for black hole evaporation, t_{evap}, found in (13) equal to the age of the universe given in (19). We then find that the mass of a black hole just terminating its evaporation at present is

$$M_{hole} \sim \alpha_G^{-1} m_p \sim 10^{15} \text{ gm.} \qquad (20)$$

The Schwarzschild radius of this black hole is just r_p, the size of a proton! (You were asked to guess this coincidence earlier.) Such a "mini hole" would end its life in a tremendous explosion. An energy equivalent to a 10^{12}-megaton bomb would be released in a matter of seconds—all from a region 10^{-13} cm in size! By witnessing such explosions, we might hope to detect these primordial black holes.

As a second corollary to "the size of the universe" coincidence, we can "explain" another well-known cosmological coincidence that concerns the number of protons in the universe. We mentioned earlier that the radius of the observable universe is roughly $R \sim ct_0$. General relativity specifies the density of the universe at a time t_0 to be

$$\rho_0 \sim \frac{1}{Gt_0^2} \sim 10^{-29} \text{ gm/cm}^{-3}. \qquad (21)$$

(Here is where dimensional arguments show their force. From the table on page 112, you can see that the combination $1/Gt^2$ is the *only* combination of G, c, t that gives a density.) Knowing the density and the size of the universe, we can deduce its mass

$$M_u \sim \rho_0 c^3 t_0^3 \sim c^3 t_0/G \sim 10^{22} M_\odot. \qquad (22)$$

We can write this mass as Nm_p, where N is the number of protons in the universe. If we now *assume* that the age of the universe is given by coincidence (19), we find a remarkable result:

$$N \sim \alpha_G^{-2} \sim 10^{80}. \qquad (23)$$

Thus, the number of protons in the universe is just the inverse square of the gravitational fine structure constant. Both this and the black hole coincidence are anthropic results if we accept that coincidence (18) is itself anthropic.

The Strong Anthropic Principle

Dicke's statement of the anthropic principle might be termed the "weak version." It says nothing about the laws of physics themselves nor anything about the actual sizes of the fundamental constants. It accepts the laws and the observed values of the constants as given and then attempts to explain several well-known features of the universe. We are now going to examine the deeper question of whether anthropic arguments can pin down the actual values of the natural constants. The notion that this may be possible is sometimes referred to as the "strong" anthropic principle. The strong principle is so called because—if true—it would obviously have much more predictive power and philosophical significance than the weak principle invoked by Dicke.

That the weak principle may not be the whole story is also suggested if we look once more at the scales shown in the diagram on page 121. The diagram reveals a curious feature: all the scales are *relative*. If the fine structure constants differed from what we observe them to be, all the scales in the figure would change but the diagram as a whole would retain the same features. For example, if G (and hence α_G) were a million times larger, planetary and stellar masses—generally about $\alpha_G^{-3/2}$—would be a billion times smaller. However, there would still be main-sequence stars, albeit with a main-sequence lifetime reduced by a factor of a million. Moreover, Dicke's argument would still apply: an observer would exist in such a universe only when its age was around 10,000 years, and he would see a universe whose mass was 10^{12} times smaller than our own! If α_G were fixed, but α was allowed to change, the effects would be less extreme, but still very noticeable. But could life arise in such a speeded-up universe? Neither conventional physics nor the weak anthropic principle has anything to say about the matter; nothing determines the actual values of α and α_G. But the strong principle must say, "No." Life can only exist if the fundamental constants have their observed values.

The first example of an argument that appeals to the strong anthropic principle was given by Brandon Carter and relates to the existence of "convective" stars. We say a star is convective when the heat generated in its core by nuclear reactions is transported to the surface primarily by way of large-scale motions of the stellar material itself. This tends to be the case for small stars (red dwarfs). By contrast, larger stars (blue giants) tend to be "radiative" in the sense that the heat gets out primarily via the flow of radiation. The dividing line between the two types is some critical mass that depends on α and α_G. This critical mass could in principle be anything, depending on the numerical values of α and α_G. But it happens to lie near the Chandrasekhar limit—that is, in the mass range in which stars actually exist—only because of the remarkable coincidence

$$\alpha_G \sim \alpha^{20}. \tag{24}$$

If α_G were slightly larger, all stars would be convective. If α_G were slightly smaller, all stars would be radiative. Now, there are arguments that suggest that planets can only form around convective stars. If we believe that life can *only* exist on planets, this means that α_G

could not be much larger than α^{20}. On the other hand, if α_G were much *smaller* than α^{20}, all stars would be chemically well-mixed as a result of convection. Such stars probably cannot form supernovae, and hence, cannot scatter the heavy elements needed for life into space. Thus, the anthropic principle already gives us one approximate relationship between α and α_G. In particular, it explains why α_G is so much smaller than α.

The fact that α_G *is* much smaller than α has another anthropic interpretation. From the diagram, the number of stars in the universe is about $\alpha_G^{-1/2}$. If we assume that the origin of life depends on chance processes that have a low *a priori* probability, then the number of potential sites for life (e.g., the number of stars) needs to be very large. This requires G to be very small, although the argument does not say how small.

We still have not pinned down the values of α and α_G; in (24) we have only found a scaling law between them. If we had one more relationship between them, we would have two equations and two unknowns, so there would exist one unique value for each number. Another relationship does exist. It does not come from an anthropic argument, but from an argument in quantum theory related to the Planck length, R_{Pl}, mentioned earlier. While the details are too complicated to go into here, the conclusion is that a self-consistent quantum field theory is possible only if

$$1/\alpha \sim ln(1/\alpha_G). \qquad (25)$$

Now we have it. Our two equations imply that α and α_G must have values comparable to what is observed. In consequence, we have now fixed *all* the mass and length scales shown in the diagram in terms of m_p and a_0. Since we always need one mass and one length to define our units, this is as good as we can hope to do.

The Weak and Strong Forces

So far we have only talked about things larger than atoms. We now direct our attention to objects smaller than atoms. On this scale, two more fundamental forces of nature come into play: the *strong* force (which holds the nucleus of an atom together and binds the quarks into an individual nucleon) and the *weak* force (which governs radioactive decay). Like gravity and electromagnetism, the strength of these forces can be described by dimensionless fine structure constants. The strong force constant is usually denoted by f^2 and has the value of 15; the weak force constant is denoted by α_w and has a value of about 10^{-10}.

By comparing f^2 and α_w to α and α_G, we see that the hierarchy of interactions is, from strongest to weakest,

strong	$f^2 = 15$
electromagnetic	$\alpha = 1/137$
weak	$\alpha_w \sim 10^{-10}$
gravitational	$\alpha_G \sim 10^{-39}$

Although both the strong and weak forces are orders of magnitude stronger than the gravitational force, they both have a very short-range effect. The strong force becomes negligible outside a distance comparable to the size of a proton, $r_p \sim 10^{-13}$ cm. The weak force becomes negligible at an even smaller distance, around 10^{-15} cm. Because the strong and weak forces are so very short-range, they do not play an important role in determining the structure of objects larger than atoms. On the other hand, gravity and the electromagnetic force are long-range forces, so it is left to them to determine the structure of large-scale objects.

The numbers f^2 and α_w are also involved in several coincidences, some of which involve the masses of various elementary particles. For example:

$$f^2/\alpha \sim m_p/m_e; \qquad m_n - m_p \sim 2m_e \qquad (26)$$

where m_n is the mass of the neutron. In words, the ratio of the strong fine structure constant to the electric fine structure constant is very close to the ratio of the proton mass to the electron mass. Also, the difference between the mass of the neutron and the mass of the proton is very nearly twice the electron mass.

Brandon Carter, who gave us the "convective star" argument, ascribes anthropic significance to these coincidences. For instance, if the mass of the proton were slightly larger than the mass of the neutron (as opposed to what we observe), then the proton would be unstable and hydrogen would not exist. The striking thing is that this and the other nuclear coincidences seem necessary for chemistry to work and to produce the elements vital for life. Indeed, given that α is anthropically determined, the nuclear anthropic conditions specify the order of magnitude of not only f^2, but of most of the elementary particle masses!

The weak force does not—at first sight—play a very important role in everyday life. And yet, the weak fine structure constant, α_w, is also involved in a striking anthropic relationship: α_w is roughly the quarter-power of the gravitational fine structure constant, or

$$\alpha_G \sim \alpha_w{}^4. \tag{27}$$

Just another coincidence? Perhaps, but this relationship is just what is required to produce an interesting amount of helium 3 minutes after the big bang, a time when the universe was hot and dense enough for nuclear reactions to occur. If α_w were slightly smaller, the entire universe would have burned to helium and there would never have been any water—a possible prerequisite for life. If α_w were slightly larger, we would have *no* helium in the universe.

Although a universe lacking in helium might not be incompatible with life, the same relationship between α_w and α_G might limit the size of α_w in *both* directions. That is, $\alpha_G \sim \alpha_w{}^4$ may explain why stars can explode into supernovae! And, as we have seen, supernovae play a crucial role in producing the elements necessary for life. If we accept (27) as a necessary condition for life, and if we believe α_G is determined anthropically, we must also accept that α_w is also determined this way.

There are even more arguments along these lines that we, unfortunately, cannot go into. To summarize them is simple: nearly *all* the constants of nature may be determined by the anthropic principle. Man, as Protagoras foresaw, has indeed become the measure of all things.

Parallel Universes

What are we to make of all this? Are we to be impressed with the anthropic principle's explanation of the aforementioned coincidences, or are we to discard it as metaphysical nonsense? Let us start by listing the objections. First, all the evidence is after the fact. It would be much more impressive if the anthropic principle could be used to *predict* a coincidence. So far this has not been done. Second, we may have been unduly anthropocentric in our point of view. We have assumed the necessity of elements heavier than hydrogen and special types of stars and planets. We have not taken into account the possibility of more exotic life-forms, such as Fred Hoyle's " black cloud." In order for a " black cloud " to be consistent with our anthropic arguments, we would have to show that its chemistry and environment required the same values of the fundamental constants as we deduced above. Otherwise, the anthropic principle as we stated it would have to be extended. Finally, the anthropic principle does not give *exact* values for the constants, but only their orders of magnitude. The situation would be more satisfactory if their values could be pinned down more precisely.

Nonetheless, it cannot be denied that there are a number of remarkable coincidences in nature, and these do warrant some sort of explanation. The point is not that there are coincidences, but that these coincidences are just what is required for life. It is this deeper level of coincidence that makes the anthropic principle so striking. Apart from Dirac's suggestion concerning the variability of G, the anthropic principle is, at present, the *only* explanation. Furthermore, every additional coincidence that can be explained by the anthropic principle does, in some sense, increase the after-the-fact evidence for it.

There remains the interpretation of the anthropic principle. From the start we admit that most physicists would be reluctant to take its metaphysical concepts too seriously. On the other hand, in grappling with the problems of consciousness, one is pushing physics to its limits. And when one pushes physics to its limits, it is probably inevitable that one will encounter a border where physics merges with metaphysics. It is at this border we find the anthropic principle. It is no wonder that most physicists are reluctant to approach this frontier. But more and more distinguished researchers are joining the ranks of those who are willing to consider the anthropic principle seriously. Not only Robert Dicke and Brandon Carter, both highly respected relativists, but Dennis Sciama, who has fathered many ideas as well as the careers of Roger Penrose and Stephen Hawking, has expressed support for the anthropic principle. Freeman Dyson, in the moving chapter "The Argument from Design," from his book *Disturbing the Universe*, writes:

> I conclude from these accidents of physics and astronomy that the universe is an unexpectedly hospitable place for living creatures to make their homes in. Being a scientist trained in the habits of thought and language of the twentieth century, rather than the eighteenth, I do not claim that the architecture of the universe proves the existence of God. I claim only that the architecture of the universe is consistent with the hypothesis that mind plays an essential role in its functioning.

John Wheeler, who is responsible for many advances in the physics of this century, perhaps the least of which was the introduction of the term "black hole" into the English language, holds a similar, if not more extreme view. Recall, we have not claimed the universe does not exist if we are not here to observe it; we have only stated that, *if* we *are* here to observe it, the universe must be the way we see it. Wheeler, on the other hand, has suggested a much more radical interpretation. In Wheeler's picture, the universe cannot come into being in a well-defined way unless an observer is produced who can eventually observe it. In this case, the existence of the universe itself depends upon life.

Although many find Wheeler's solipsistic standpoint unpalatable in view of its rather mystical overtones, there is another framework for the anthropic principle which might seem more plausible. This is the "many worlds" interpretation of quantum mechanics, proposed nearly 20 years ago by Hugh Everett. One of the underlying features of quantum theory is that a system is not in a well-defined state until one makes a measurement or observation on it. Prior to the measurement, the system—which might in principle be the entire universe—is in an undefined, fuzzy state. It is not in one state or another but in a sort of combination of states, what we call a *superposition* of states. It takes an act of measurement or observation to force the system into a particular one. Quantum mechanics allows one to predict the probability that the measurement will have a specific outcome, but it does not determine the result with certainty.

This concept is very strange and leads to apparent paradoxes in some situations. The most famous of these is the "Schrödinger cat paradox." A cat is imprisoned in a box. Also in the box is an atom of a radioactive element with a half-life of one hour, a geiger counter and a vial of poisonous gas. If the geiger counter finds that the atom decays during the hour of the experiment, it releases the gas and kills the cat; otherwise, the cat survives. By the definition of half-life, after one hour there is exactly a fifty-fifty chance that the cat will have been killed. Now, in quantum mechanics, the equation that describes the state of the cat at the end of the hour only does so in a fuzzy way. It says that, until someone looks inside the box, the cat is made of fifty percent live cat and fifty percent dead cat. Everett resolves this paradox by saying the universe splits in two. In one universe there is a live cat, in the other a dead cat.

The same idea can be extended to any other observations, even when there are more than two outcomes. One envisages a universe that splits whenever an observation is made. Each split corresponds to a possible outcome of the observation. The universe is continually branching: cats live, cats die, wars break out, wars are averted. All worlds differ by a decision or a cat. Perhaps, in addition, some worlds differ in the values of their fundamental constants. In some universes α is big, in others, α is small. Only in a tiny fraction of these many worlds will the values be such that life can evolve. On these worlds intelligence flowers and becomes aware of itself. On these worlds men may develop an anthropic principle to explain their existence. On other worlds, life will never evolve and the questions will never be asked.

6

The Peryton and the Ants

by E. C. G. Sudarshan & Tony Rothman

In many ways, this article is the most personal of the series and, as such, deserves, the most personal introduction. It attempts both to explain certain basic physical phenomena and to relate them to the physicist's view of the world. Upon rereading the original 1979 version of the article, I felt that it partly succeeded in the latter but that the technical discussions were too convoluted to hold the reader's attention. I have therefore made some revisions for this volume which I hope will make the going a little easier. Quantum mechanics is not an easy subject.

The article grew out of long hours of freewheeling conversation with a very unusual physicist indeed, George Sudarshan. Professor Sudarshan is the director of the Center for Particle Theory at the University of Texas in Austin, which is located on the same floor as the Center for Relativity. He is a fellow of the American Physical Society and the Indian Academy of Sciences and is the author of several graduate-level texts as well as 300 scientific papers. Among his discoveries are the V-A theory (the first theory of the weak nuclear force) and the optical equivalence theorem (which is the connection between classical and quantum optics). It is not surprising that Fritjof Capra acknowledges Sudarshan in The Tao of Physics. *Contrary to what you may have heard elsewhere, it was Sudarshan who first conceived the hypothetical particles called "tachyons" (*American Journal of Physics, *1968).*

This article is printed here for the first time.

In the past, people have tended to view the study of the physical world as being distinct from the study of the social world. Most people who studied one did not find the other interesting, or, if they did, would never admit it in print. Yet, just as it is a sign of maturity in individuals that they show an interest in both the world that surrounds them and in the society in which they live, so it is a sign of a civilization's maturity when it breaks down the distinction between the study of the two worlds.

Physics is the science that has traditionally been viewed as the farthest removed from personal experience. It deals with such dry concepts as matter, position, acceleration and velocity, which seem to have little in common with society and even less to do with romance, literature or film. With the development of solid-state circuitry, pacemakers, microwave ovens, nuclear weapons . . . ad infinitum, it is now commonly acknowledged that advances in physics have come to permeate our daily lives. Nonetheless, it is somehow still felt that the underlying physics is very far removed from the ability to enjoy sex or a hot cup of coffee. We are going to argue that there is a way of viewing physics that makes them seem less distantly related. We are going to make this argument in the context of quantum mechanics.

Quantum mechanics and relativity are often called the two towering peaks of twentieth-century physics. Straining the metaphor a bit past the elastic limit, we see that relativity, like the sacred Mt. Fuji, has attracted the pilgrimage of millions because of the mystic appeal of quasars, pulsars, black holes and wormholes. Quantum mechanics has not enjoyed such a warm glow of publicity. Indeed, tourists have avoided it as the Scylla of physics, whose treacherous and uncertain waters doom all but a few hardy voyagers. Popularizers, for their part, have little trouble explaining that a black hole is something you can't get out of once you get into (much like being in debt, we imagine), but they uniformly fail to conjure a picture of the quantum mechanical wave function. Admittedly, sometime around the fifth or sixth grade we all hear that if you pound your head against a wall for a billion years or so, there is a small chance you will actually pass right *through* the wall, unscathed, and your pain will be over. (We don't recommend this procedure; the odds of success after even a billion years are still quite small.) When asked how this can possibly occur—without the

133

wall or our skulls breaking—we are told it has something to do with quantum mechanics. That is the extent of it. And thus, the average citizen grows up believing Pope's verse:

> Nature and Nature's Laws lay hid in night:
> God said, Let Newton be! and all was light,

with perhaps a small correction term due to Einstein.

Newton's attempts at explaining the world were very modest. His physics dealt only with the motion of inanimate objects. In one sweeping gesture, his laws of motion united the movement of apples, moons, suns, planets and tides. His laws, like the Ten Commandments, have passed into our common inheritance: *Any object in motion will continue moving in a straight line at constant speed unless acted upon by an outside force.* And all the people saw the thunderings and the lightnings and heard the noise of the trumpet . . . .

And so, we become adults in a Newtonian world: we see cars colliding with one another, like so many billiard balls; we view our families and neighbors as distinct entities; we add two apples and two apples to get four apples; we know that apples do not spontaneously change to oranges—in fact, nothing we can do changes apples to oranges.

Indeed, the Newtonian world is essentially a billiard-ball world. To be a proper Newtonian physicist, you must sit back in your chair and pretend *everything* is a billiard ball. Moons, planets and suns are surely nothing if not oversized billiard balls. Cars can be approximated as billiard balls, and, if they sometimes stick together after colliding, it is only because of forces between smaller, invisible billiard balls. (As a matter of fact, a recent Nobel Prize winner, Ilya Prigogine, once modeled a traffic flow pattern as a problem in hydrodynamics—the flow of an apparent continuum made of tiny billiard balls down a tube.)

People, too, in this Newtonian world might as well be billiard balls. Tony is the 8-ball. George, sitting to his right, is the 10-ball. Just as Newton held that all interactions took place between separate, billiard-ball objects, we also view each other as distinct, corporeal, immutable entities. Although the marriage vow we intone says, "Until death do us part," we are, in fact, almost never together. Interactions between people usually seem to be of the collisional type—words or fists bouncing off one another. We often feel great swells of jealousy when someone we love shows interest in another. But at the same time, how frequently, when the competition is absent, do we make ourselves available to our loved one? How often do we experience an intense feeling of love—the feeling of inseparability from our companion? This raises a paradox unresolvable in Newton's world: How can you be inseparable from another and yet retain your own identity? If this sounds too unscientific, let us rephrase the question: How is one to observe a system without interacting with that system?

Suppose now that you are on the verge of falling asleep. Things become a little fuzzy. If you have poor eyesight, so much the better; take off your glasses. The letters on the headline of the newspaper become indistinct, impossible to separate from one another. The apple and the orange sitting on top of the counter have blurred into one. Even your two friends—previously distinct entities—have joined to become a sort of Siamese twin. Cars colliding with each other on the freeway do not seem so much like billiard balls any more. They are less localized—spread out in a blur that oozes about, thinning and thickening, depending on where the traffic is the heaviest.

When we are in such a state of consciousness, our first reaction is usually an attempt to get out of it. Something, we decide, is wrong. Our glasses are off so we're not seeing straight, or we're drowsy so we're not thinking straight. We put on our glasses or splash water in our faces. Then we can start thinking again; then we can be proper scientific observers. Anyone who wants to be a scientific observer must be alert and awake so that he doesn't see a vague, dreamy blur of billiard balls that have no definite place. He must see the balls as being definitely here or there, or flying apart in precise trajectories. Science exists in the state of wakefulness.

But would it not be interesting to make a science in the dreaming state? Then people could fly and trees could talk and billiard balls would be like puffs of smoke. Solid objects would become elusive like whiffs of perfume, but the surrounding atmosphere would become tangible and clouds would make hard walls. Could we practice science in such a world?

If we could, it would be a strange new type of science. The elementary building blocks would have no definite position, no definite state of motion. Situations would be such that we could only say that this billiard ball looks mostly red, though it is also a little blue—not an intermediate color, nor an Easter-egg pattern, but sometimes red and sometimes blue. It is almost like trying to decide the color of a loved one's eyes! In this science, elephants would ooze through walls like steam through a garment that is being ironed. Clouds could become hard, localized objects that you could catch in the palm of your hand. There would still be a dynamics—a theory of interaction—to this crazy-quilt world. It would have the logic of a dream; persuasive and natural, but not one that makes too much sense after you have passed the twilight of waking and have sipped a cup of morning coffee. The more you tried to elucidate it, the less sensible it would sound. Perhaps you would be forced to discuss it in a new language of symbols and manipulations—somewhat like what young Joseph did for the Pharaoh, or Freudian psychoanalysts do for our neuroses. It might borrow from poetry, sculpture or modern mathematics. It would make perfect sense in its context, but only there. To participate, you must alter yourself and "become like a child." As Kekulé would have said, "Gentlemen, let us learn to dream!"

But such a science already exists! It is called quantum mechanics. It insinuated itself among the sober gentry who were eternally vigilant at their physics near the turn of the century; but it came into its own in the 1920s through the work of Erwin Schrödinger and Werner Heisenberg, both of whom must have been less than sober to have proposed such theories as serious physics of the subatomic world.

We are now going to advocate that the Newtonian, billiard-ball world described earlier is perhaps, in fact, not the world we experience daily. We are soberly going to champion "fuzzy physics" in the hope of persuading you that the world we experience daily corresponds more closely to the wistful, quantum mechanical world of oozing elephants and solid clouds.

The path of least resistance in winning an audience over to a strange idea is always to use the ploy of "all knowledge is rearrangement" (much in the spirit of Plato's "all knowledge is remembrance"). That is, we assure the audience that what we have to say is not very strange at all—that it has been thought of a million times before and that we are just saying it differently. We may take the beautiful passage from the Presocratic philosopher Melissus (fl. 440 B.C.), as an example relevant to our discussion. The language is a little difficult but worth struggling through. You should hear strong resonances of what you have read a few paragraphs earlier.

If there were a plurality, things would have to be of the same kind as I say the one is. For if there is earth and water and air and fire, and iron and gold, and if one thing is living and another dead, and if things are black and white and all that men say they really are—if that is so, and if we see and hear aright, each one of these must be as we first decided, (that is) they cannot be changed or altered, but each must be always just as it is (from the start) . . . .

And yet we believe that what is warm becomes cold, and what is cold warm; that what is hard turns soft, and what is soft hard; that what is living dies, and that things are born from what lives not; and that all those things are changed, and that what they were and what they are now are in no way alike. We think that iron, which is hard, is rubbed away by contact with the finger; and so with gold and stone and everything which we fancy to be strong, and that earth and stone are made out of water; so that it turns out that we neither see nor know realities.

Now (the two arguments) do not agree with one another. We said that there were many things that were eternal and had forms and strength of their own, and yet we fancy they all suffer alteration, and that they change from what we see each time. It is clear, then, that we did not see aright after all, nor are we right in believing that all these things are many. They would not change if they were real, but each thing would be just what we believed it to be; for nothing is stronger than true reality. But if it has changed, what is has passed away and what is not has come into being. So then, if there were a plurality, things would have to be of just the same nature as the one (see Reference 1 at the end of article).

In everyday life this is so: the food we eat becomes part of us; parts of us such as hair, nails or sweat cease to be attached to us. Food transforms the baby into the youth, who in turn becomes the adult, and the dead body of the old person merges with the elements. " Dust thou art, unto dust thou returnest." The breath of life (or the genetic information, if you want to be scientific) plays with matter, generating the growing, active being, and, after having tired of the game, throws the toys of chemical elements and compounds into the environment.

All of chemistry and technology is in the same mold, too. All processes are simply rearrangements of the basic building blocks; in chemistry, the atoms are the blocks and chemical reactions are like square dances: " Swing your partner round the square, round and round the room you dance; change your partner to the forward side . . . ." Technology makes use of the laws of physics and simply directs the natural processes to bring about desirable ends. Commerce is no different. We will find that " the many is the same as the one " is just as true in quantum mechanics as in our common experience.

We would now like to engage in a little taxonomy in order to contrast three world views. First we will deal with Newtonian particle mechanics (or *classical* particle mechanics, as it is called equally often) and classical wave mechanics. Later in this article quantum mechanics will make its appearance, by which time the parallels with everyday life should be becoming apparent.

We have already discussed Newtonian particle mechanics to some extent. When building a Newtonian model, we tend to divide up the world into discrete objects (billiard balls). These billiard balls are permanent and immutable and are the fundamental reality of the world. If we do not perceive these billiard balls directly, then we aren't perceiving reality. Take, for example, a box of gas or, better yet, a cloud. What you see is a large fluffy object with indistinct boundaries. Newton says this is not what is real. What are real are the tiny, invisible billiard balls bouncing around inside the cloud. The physics describes the motion of these billiard balls and, since it is the billiard balls with which the physics is concerned, they are what is real. So, in a sense, Newtonian mechanics asserts that clouds are part of a hallucination. If you do not directly perceive billiard balls, you are under a delusion.

Moreover, in Newtonian mechanics there is a distinct separation between substance and process. The motion of the billiard ball is not the billiard ball itself. This is clear. Although some people advocate, "You are what you eat," we have yet to hear anyone claim, "You are your walking."

After it was developed, the Newtonian picture proved so successful that it was used even where it didn't apply. We are speaking of hydrodynamics. It is rather remarkable that it should occur to anyone to view honey oozing out of a bottle as a collection of tiny, invisible billiard balls. Yet, this is exactly what hydrodynamics does: It postulates that the bulk properties of a liquid motion can be deduced from the movement of microscopic constituents. We can thus think of hydrodynamics flow as the correlated motion of many particles moving in one direction.

Let us now divert our attention to waves, stressing that we are talking about classical waves and have never heard the term "quantum mechanics." Waves are among the most common phenomena we experience: pebbles dropped into water make waves; radio waves permeate (and often pollute) the atmosphere; light travels by waves; sailplanes use airwaves over hills to find altitude. The wave view of the world is, in some ways, quite different from the billiard ball view of the world. It is a permissive, interpenetrating structure in contrast to the uptight, exclusive structure of billiard balls.

First, let us ask, "What is a wave?" Perhaps it is easier to ask, "What is a wave not?" Consider the standard and fairly dull example of a cork floating on the surface of a pond as a wave comes by. If you are looking down at the cork from the bank of the pond, you swear to yourself that a wave is traveling along and hitting the cork. This is to say, all of us swear that the water moves along, hits the cork and continues onward. This is not true. If you stepped into the pond and saw the cork at eye level, you would discover the only thing that happens when the wave comes by is that the cork bobs up and down. There is no lateral movement of the cork—something you would expect if the water were traveling along. Contrast this behavior with the hydrodynamic scenario. Honey oozing out of a bottle will carry along a fly who has gotten caught in the flow. The concerted motion of the honey molecules, all in one direction, is not a wave.

The behavior of the cork in the pond is exactly analogous to the illusion often seen on movie marquees. The lights flash and the design or pattern of lights travels around the edge of the marquee. But, as you realize after fixing your attention on a single bulb, the individual lamps are simply flashing on and off in their places. The rapid sequence of flashes creates the illusion of the moving lamps. The design *is* moving; it is shifting position from one set of lights to the next. The lamps are not moving; the design is not the lamps. Similarly, if we looked closely enough at the water molecules, we would find no lateral motion, only up and down motion as the wave came by.

Here we have a good example of a case where the medium is not the message. The lamps on the marquee are not a wave; the water is not a wave. Extending this argument to the case of sound, it would be foolish to call the jiggling of air the wave, because the same wave causes our eardrums to jiggle, too (if it didn't, we would not hear the sound). What is a wave, then? Well, the wave is the design on the marquee or the energy transmitted through the water or the air or our eardrums. When we study light, we see that it is transmitted not only through air and water, but also through the vacuum of space. Here we drop the medium altogether and say that the message is transmitted through nothing.

As long as we are on the subject of waves, this is a good time to introduce some terminology which will be very useful later. The rate at which the water molecules or the cork bob up and down is called the *frequency* of the wave. It is usually denoted by the Greek letter v (nu) and is measured in cycles per second (which has been designated Hertz, just to obscure the meaning). Frequency can also be thought of as the number of wave crests that pass a given observation point each second. The distance between two successive peaks of a wave is called the *wavelength* and is usually denoted by the Greek letter λ (lambda). The wavelength and frequency are related by the simple equation

$$\lambda = \frac{c}{v},$$

where c is the velocity of the wave. If c is assumed known, then it is clear that by knowing the frequency of the wave, we know the wavelength, and vice versa. Keep in mind that as the frequency goes up, the wavelength goes down, and again vice versa.

We have hinted that a wave and a billiard ball are two very different animals. It is very easy to point to the corner pocket and say, "There is a billiard ball." It is not so easy to point to a pond and say, "The wave is right *there*." If you did, we would shoot back with, "And where is *that*? Do you mean that crest or that trough or the whole bunch of them—or what?" A wave is nonlocalizable. You can't pinpoint its position; it is much like a politician. That you can't pinpoint its position is quite understandable; a wave is more like a process itself, and it is very difficult to pinpoint the position of a process.

The situation is even fuzzier when we consider more than one unit. It is easy to say, "There lie two billiard balls. Count them: one, two." One plus one always makes two in the Newtonian picture. Not so with waves. When two water waves of equal height or two light waves of equal height collide (or *interfere*), several interesting things happen. First, the two individual waves combine (or *superpose*) to form one wave with a height equal to twice that of the initial waves. They have lost their individual identity to produce a third wave. There seem to be shades of Melissus lurking here. Secondly, the two waves pass right through each other without disturbing one another. Laser beams generally do not bounce off one another when crossed. Thus, the waves add but do not disturb. Interference without interference—a strange state of affairs.

Well, you might object that since the waves pass undisturbed right through each other and yet add up to to twice their original heights (or *amplitudes*), can't we say, "This is what we *mean* by two waves being over *there*? After all, we can decompose the big wave into the two separate smaller waves again, can't we?" Here you would be correct. We can always take a wave and decompose it into two smaller waves, each of one-half the original amplitude, or three waves of one-third the original amplitude, or a million waves of one-millionth the original amplitude. Decomposition is just the inverse of superposition.

But we must be careful not to be tempted into thinking of waves as extended billiard balls. We may have been giving the impression that amplitudes are very Newtonian, that they add as billiard balls add. This is true only if the two waves add together *in phase*. This means that the crest of one falls exactly on the crest of another. Then, and only then, do amplitudes add like billiard balls. However, if the crest of one should fall on the trough of another—that is, if the waves are *out of phase*—the amplitudes add up to zero. No two billiard balls ever added together to get zero. And, of course, you cannot take a pile of 10 billiard balls and divide it into a pile of $7\frac{1}{2}$ balls and $2\frac{1}{2}$ balls. But you can divide a wave with amplitude 10 centimeters into two waves of amplitudes $7\frac{1}{2}$ and $2\frac{1}{2}$ centimeters. Thus, you can see, it is "no go" to think of waves—or even their amplitudes—in terms of billiard balls.

There is an even more serious problem with thinking about waves in Newtonian terms. Take two identical laser beams. If we add the beams together, we see portions of the beam, where the two laser beams are in phase, that are not twice as bright as the separate beams, but *four* times as bright. Along similar lines, if you double the voltage to a light bulb, you quadruple the output. (This explains why light bulbs tend to burn out in power surges.) Here we are talking about the *intensity* of a wave. The intensity is a measure of the wave's energy. While in physics we have both amplitudes and intensities, it turns out that normally we observe the latter. (And intensities do not add linearly, as do amplitudes, but as the square of the sum of the amplitudes.) The human eye perceives intensities; this is why part of the resulting beam from the two lasers appeared four times as bright as the one. We see, then, that in many ways the whole is far greater than the sum of its parts. It has recently become fashionable to call effects that add in phase "synergistic," though the word is at least three centuries old. And the concept is much older. In fact, we may liken this state of affairs to being in love.

Consider an ant. Or rather, let us defer to Lewis Thomas, who has considered ants many times before:

A solitary ant, afield, cannot be considered to have much of anything on his mind; indeed, with only a few neurons strung together by fibers, he can't be imagined to have a mind at all, much less a thought. He is more like a ganglion on legs. Four ants together, or ten, encircling a dead moth on a path, begin to look more like an idea. They fumble and shove, gradually move toward the Hill, but as though by blind chance. It is only when you watch the dense mass of thousands of ants, crowded together around the Hill, blackening the ground, that you begin to see the whole beast, and now you observe it thinking, planning, calculating. It is an intelligence, a kind of live computer, with crawling bits for its wits (ref. 2).

Let us analyze this colony from the various points of view we have so far discussed. A die-hard Newtonian would say, "Here we have an ant colony that is composed of a million particles—the ants—all following different trajectories and engaging in collisions with one or more other ants."

If we were thinking in terms of traffic flow, we might say, "Here we have the hydrodynamic scenario: many ants acting in concert. This is like many honey particles participating in fluid flow, or many auto-mobiles traveling down the street."

Abandoning the billiard-ball approach, one could say, "Here, the ants don't seem like particles at all. Rather, they are like small wave-lets superposing to produce a big wave with an amplitude of 1 million ants. Moreover, just as the two laser beams produced a resultant intensity greater than that of the sum of the two beams, the ant colony has an intelligence greater than that of a million individual ants."

The ant colony is of interest because it allows for description in terms of several models. Depending on our frame of mind, we can either view the Hill as a collection of particles, a million billiard-ball ants or as a large wave. In this latter description, we again hear echoes of Melissus: While the colony as a whole can be characterized as being composed of a million ants, the million ants must be thought of as one entity, the colony. In order not to prejudice you, we leave it to you to decide which model comes closest to describing your daily life.

You might object that we are just playing with words: the ant colony is still the ant colony no matter how we want to describe it, and our various descriptions do nothing to alter the fact that it is neither a collection of particles, nor a wave, but an ant colony. In response, we would argue, that reality is what we perceive, so that what is real depends on the model in which we are working. To make this point clearer, pretend that something existed that really was *both* a wave and a particle—not either, but both. How would we perceive it?

As a graphic example of this type of animal, consider the mythical beast, the Peryton. According to Borges, "The Perytons had their original dwelling in Atlantis and are half deer, half bird. They have the deer's head and legs. As for its body, it is perfectly avian, with corresponding wings and plumage. Its strangest trait is that, when the sun strikes it, instead of casting its own shadow, it casts the shadow of a man . . ." (ref. 3). For our purposes, it will be necessary to alter this description a bit. Pretend that we can *only* observe the Peryton by catching its shadow. Pretend further that, instead of casting a shadow of a man, it casts a shadow alternately of a bird or a deer, depending on its mood. Now we have the perfect example of a quantum mecha-nical wave packet. We cannot call the Peryton a bird or a deer, and we cannot say it is half-and-half either. Since in our revised version of the myth we only observe the Peryton by its shadow, we can only say that sometimes it presents itself as a bird and sometimes as a deer. But it is not half-and-half. If it were, we would see a shadow of half-bird, half-deer. We can say only that the Peryton is *both* deer and bird.

In the case of a quantum mechanical wave packet, we say that it has the characteristics of both a particle and a wave. Which "shadow is cast" depends upon the situation. Sometimes the wave packet is localized and hard, like solidified tar, which could be rolled up into a billiard ball and smacked with a cue stick. At other times the wave packet is soft and nonlocalizable, like melted tar on a warm day. Perhaps you would prefer an analogy with a cloud: sometimes we speak of a cloud as being one small, compact object—*the* cloud; other times we speak of it as being like fog and hard to pin down.

The important point to remember when thinking about our Peryton or about wave packets is that you cannot observe all properties of the object simultaneously. You see the bird now, the deer later; the particle today, the wave tomorrow. Yet, experience cannot be self-contradictory; thus all the observed properties must be characteristic of the same object. We conclude, as we have said before, that the Peryton must have properties of both the deer and the bird, and that the wave packet must have the properties of both a wave and a particle.

There are several famous experiments that clearly illustrate the "dual nature" of wave packets and that deserve to be treated at some length, both to initiate the uninitiated and to correct misimpressions shared even by some scientists.

It is well known that light waves falling on a metal surface will cause electrons to be ejected from this surface. These electrons can be coaxed into an electrical circuit, where they form an electric current. This ejection of electrons by light is aptly called the photoelectric effect and finds its way into many aspects of our daily lives. It is the operating principle behind solar cells, not to mention automatic door openers and highway car counters.

The photoelectric effect was very puzzling to physicists at the turn of the century. Keep in mind that the light incident on the metal surface is in the form of a wave. It can be focused with a lens or reflected by a mirror and has many standard wave properties. For instance, its wavelength and frequency can be measured in exactly the same way as was done for water in the pond. The incident light is characterized by an amplitude and the square of this amplitude, the intensity. As we mentioned before, intensity is a measure of the beam's energy: the brighter—or more intense—the beam is, the more energy it has.

Now let us examine the behavior of the emitted electrons as we do the experiment. First, we find that *no* electrons at all are emitted until the frequency of the light goes above a certain threshold value. Once the threshold is reached, we find that the energy of the individual electron ejected depends on the frequency of the light. That is, if the frequency of the light is doubled, the energy of the electron is also doubled (ignoring the small contribution of the threshold frequency). On the other hand, we observe that if the *intensity* of the beam is doubled, the energy of each electron *does not change*. What does change with intensity is the number of emitted electrons. Double the intensity and you double the number of electrons given off.

It is very difficult, in fact impossible, to explain the above experimental results in terms of the wave nature of light. Why is this? We know that intensity is a measure of the energy of a wave. Double the intensity and you double the energy. Note also that intensity, the square of the amplitude, has nothing whatsoever to do with frequency, the rate of oscillation. So, *regardless of the frequency*, we could make the wave as energetic as we wanted by increasing the intensity. Surely then, at some point an electron in the metal would receive enough energy to be knocked out. Yet we find just the opposite to be true: light of too low a frequency will *never* eject an electron, no matter how intense the beam. To make matters even more puzzling, once the threshold frequency is exceeded, increasing the intensity still does not affect the energy of the individual electrons; only the number of electrons emitted is affected.

Einstein concluded that the only explanation for this behavior was that light sometimes acts as particles, or quanta, called photons. He wrote a very simple relationship between the energy of the electrons and the frequency of the incident light:

$$E_{\text{electron}} = h\nu - E_{\text{threshold}}.$$

Here, h is a number called Planck's constant. It is just a fixed unit of action; that is, energy multiplied by time.*

This equation explains the state of affairs very nicely. The quantity $h\nu$ can be thought of as the energy of one quantum of light, one photon. It is the energy that the photon must give to the electron to raise it above the threshold energy and impart to it an extra energy, E_{electron}. We see that if $h\nu$ is below $E_{\text{threshold}}$, the electron will not be ejected, as was observed before. Notice that E_{electron} depends only on the frequency of the photon, a point that was emphasized earlier. But, in this picture, the intensity of the beam finds a natural interpretation: the number of photons incident on the metal. If we accept that each photon knocks out one electron, we see that by doubling the intensity

---

* In "Coincidences in Nature," we used the symbol $\hbar$, which was more convenient. Technically, h is Planck's constant and $\hbar = h/2\pi$.

of the beam, we double the number of incident photons, and thus double the number of exiting electrons. This we also found to be true.

Generalizing from the photoelectric effect, we postulate that the energy of any photon is given by $E = h\nu$. We also assume that the momentum of a photon is given by $p = h\nu/c$, a relationship that will be used shortly.

This entire situation is quite curious. The incident wave was spread out. The wavelength of visible light is on the order of 100 times the diameter of an atom, so we expect all the energy to be spread out as well. Yet, all the energy of the wave seemed to be gathered together by the electron as it emerged from the metal. Even more curious is the relationship between the energy of the electron and the frequency of light, the former being manifested in the forward motion of the electron and the latter being associated with something that moves up and down.

You should not feel guilty if you are uncomfortable with the preceding discussion. For the above analysis, Einstein was awarded the Nobel Prize in 1921, and he is supposed to have said, "Anyone who claims to understand $E = h\nu$ is not being entirely truthful."

The correctness of the above picture was dramatically demonstrated in 1922 by Arthur Compton. He systematically observed the scattering of X rays from a graphite target and made a plot of the energy of the scattered wave. Compton found that his scattered X rays were of a lower frequency than his incident X rays. Since $E = h\nu$, this means the X rays lost energy.

The result greatly surprised Compton, who did not expect an energy loss at all. He was thinking in terms of a naïve theory of scattering, known as the Thompson theory, which predicted that the outgoing X rays should have the same energy or frequency as the ingoing X rays. Obviously, something was wrong and Compton could only solve the puzzle by treating the photons as billiard balls with a momentum $p = h\nu/c$.* Then, using the basic rules of conservation of energy and momentum, he was able to derive a formula that precisely explained the anomalous results.

In our discussion of the photoelectric and Compton effects, the particle properties of photons have been emphasized. There is, however, a third experiment that points out even more clearly the fact

---

* The physicist who keeps dates in mind or the historian of science will find an interesting puzzle here. At first sight, Compton seems to be using the formula $p = h/\lambda$ ($= h\nu/c$) in 1922, two years before it was postulated by de Broglie. There are several reasons why this expression is a logical choice; nonetheless, it must have been an assumption whose validity was not at all clear at the time.

that the Peryton-photon sometimes casts the shadow of a wave and, at other times, the shadow of a particle. This is the famous two-slit interferometer. The following discussion is conceptually more difficult than the previous, so you should gird up your loins and fully digest your last meal before proceeding.

Consider a laser beam that uniformly illuminates two slits cut into an otherwise opaque sheet of metal or some other nontransparent material. The light waves from the laser impinge on the slits and undergo a process called *diffraction*. We experience diffraction in ordinary situations. If we stand on a bridge and look down into the river, we often see water waves diffracted around the pylons supporting the bridge. The water waves, which initially were flowing straight down the river, get bent around the pylons and begin moving in slightly different directions. The same phenomenon holds true in the two-slit interferometer. The impinging light waves are actually bent around the edges of the slits and are redirected in a distinct manner—a manner predictable from classical wave theory. If we project these bent rays onto a screen, we get the characteristic, "two-slit interference pattern" (see Fig. 1). The bumps in Fig. 1 represent bright spots where the redirected light has constructively interfered with itself. That is, by bending the light in a new direction, the slits have caused some crests to fall on other crests and form bigger waves. At the same time, some crests now fall on troughs and at these points the waves have destructively interfered. On Fig. 1, the troughs represent places of destructive interference, where the amplitudes have added up to zero.

Now, it turns out that there exists an easy method by which we can use the distance between neighboring bright spots on the screen to measure the wavelength of the laser light. Thus, the two-slit pattern gives us direct information concerning the wave properties of light. Furthermore, it is easy to show that the light must go through *both* slits to produce the two-slit pattern. We do this by covering up one slit and discovering that the characteristic pattern has disappeared; instead we find a different, "single-slit" pattern. We can even reduce the illumination so much that we can think of one photon at a time passing through the slits. (Now we are thinking of the beam intensity as being proportional to the number of photons—the Einstein photoelectric picture.) If we wait long enough for a pattern to build up—with both slits open—we find that we still get the same two-slit pattern. We conclude that each photon has passed through both slits. (You might ask how we know that the first photon didn't go through the top slit, the second photon through the bottom slit, and so on, until the pattern was built up. The answer is that, if this were so, we

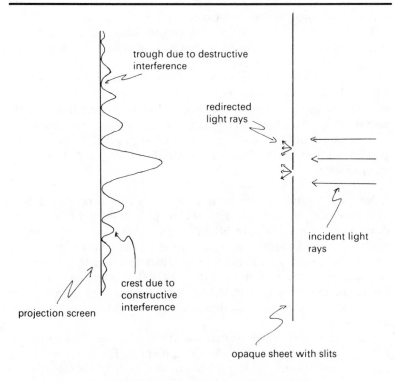

trough due to destructive
interference

redirected
light rays

incident light
rays

crest due to
constructive
interference

projection screen

opaque sheet with slits

FIGURE 1 The two-slit interferometer.

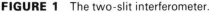

would get two single-slit patterns, which is *not* a single two-slit pattern.)

The situation is made doubly confusing when one realizes that the screen, say a photographic plate, records the presence of a photon only at a *single point*. Only *one* grain of film emulsion is blackened per photon. (Note that we have not said the photon exists only at a single point, but only that its presence is recorded at a particular point.) The final interference pattern is the collection of darkened grains, separated by regions of unexposed grains, but one must step back a little way to see how the dots fall into the stripes of light and dark that are characteristic of the two-slit pattern. We might call this effect the "pointillist interpretation of quantum mechanics," after the style of the painter Seurat. Up close, we are aware of small dots of colored paint. Stepping back further, we see *A Sunday Afternoon on the Grande Jatte*. With this point of view in mind, we say *the interference pattern is a measurement of the photons' positions on the screen.* We will have occasion to discuss the concept further.

The fact that all the energy of a photon has been collected by a single grain of film, reminds us of the photoelectric effect, where all the energy of the incident wave was collected by one escaping electron. Actually, a similar phenomenon occurs in everyday physics: a radio wave may have a wavelength of meters or hundred meters and is thus very spread out. Yet, an ordinary pocket radio manages to collect the energy at a single location. In the present two-slit experiment, the photon must travel as a spread-out wave to produce an interference pattern that can be detected as a particle at a single point. This should seem strange.

No doubt some of you remain unconvinced; there should be a method to determine from which slit the photon emerged. Surely, we can put a counter directly behind a slit to determine whether the photon passed through it. If we succeed in making this determination, we will have destroyed quantum mechanics by creating a contradictory situation: The photon went through both slits (we have an interference pattern); the photon went through one slit (we detected it at the slit with a counter). Quantum mechanics would then obviously be inconsistent.

One can, indeed, determine the slit through which the photon has passed by using a counter in the above manner. For the moment, we only note that such a determination will destroy the *interference pattern itself*, not quantum mechanics. Since the existence of the interference pattern indicated that the photon had passed through both slits, by destroying the pattern we have destroyed any contradiction. We no longer know the photon went through both slits. It is impossible to have simultaneous experience of both the pattern and knowledge of the photon's choice of slit. Not all properties of the Peryton can be observed simultaneously.

That there exist properties that cannot be measured in all circumstances is the most important message of the two-slit experiment. Before going on to other topics, however, let us reconsider the same experiment; this time the slits will be illuminated by two identical but distinct light beams,* each of half the intensity of the original source. We may put a tiny photoelectric counter at any point behind, in front of or near the slits to record the passage of photons past that point. We call this a local measurement. Now, any measurement by the counter will record exactly the same number of photons incident on

---

* We are excluding such systems as phase-locked lasers.

any point as before when we used one light source. The flux of photons (illumination) has not changed. Thus, by local measurement, we cannot distinguish between the illumination caused by one light source and that caused by two light sources. However, when using two light sources, we will find that the interference pattern on the screen has vanished—absolutely, completely. There is some nonlocal property of light that the counter has failed to detect. It is a property that cannot be measured right at the slits but that takes time to develop and manifest itself as a pattern on the screen. This property is called *mutual coherence*. It is a way of sharing a photon between two locations. A greater coherence implies a greater sharing of photons. The two independent light sources have no mutual coherence, the photons are not shared by the slits, and no two-slit pattern is produced.

Clearly, if the beams are perfectly coherent, the photon is shared equally between the two slits. It has passed through both. Also, the greater the coherence, the sharper the interference pattern on the screen. At this point, we might guess that each of these states of affairs implies the others. That is, to say that the photons passed through both slits is equivalent to saying both that the two beams were mutually coherent and that an interference pattern can be produced. All this is true.

On the other hand, by putting the tiny counter behind one slit or the other, we can decide whether a photon passed through that particular slit. We now realize that a local measurement of illumination (as we have been using the term) is equivalent to the ability to determine through which slit the photon has passed. But we previously stated that by using a local measurement we forego any knowledge of coherence; hence, by using a local measurement, we forego any knowledge of the interference pattern.

Coherence is a property that we often encounter. The term synergy has already been mentioned, as has love. Here is also a fourth way to describe the behavior of the ant colony: two ants alone are not acting very coherently, yet many ants together are acting as if their intelligence is shared between all of them. But notice, as the mutual coherence of the ants increases, it makes less and less sense to talk about them as individual ants. We are not only saying that two heads are better than one, but that two heads *are* one. Perhaps intelligence is not linked to consciousness but to coherence.

You can either talk about individual, incoherent ants, or you can talk about one coherent ant colony. You cannot have your cake and eat it, too—as our Jewish mothers used to say. The incompatibility of measuring two quantities is perhaps hard to swallow; certainly in nineteenth-century literature, the omniscient author assumed he knew everything that transpired in every character's head. Only in this century have writers realized the contradiction of entering the minds of both Andrey and Natasha at the same time. When mind-melders are invented, mutual coherence will increase, but identity will be lost.

In physics, the most famous example of this incompatibility is the impossibility of obtaining precise measurements of both the position and wavelength of a wave. Visualize a complicated wave such as that drawn in Fig. 2. In order to get a good fix on the wavelength, you need to measure many cycles to be sure you have gotten a full wavelength and not just the distance between two intermediate bumps. But if you measure many cycles, you make the position uncertain. Remember how much trouble we had in pinpointing the position of water waves spread over a pond. We can make the position measurement more exact by considering only a portion of the wave, but then we are unsure if we have correctly measured the wavelength. The best we can do is measure the position and the wavelength such that

$$\Delta x \Delta k \sim 1.$$

(Read, " Delta x times delta k is of order 1.") Here, we have used for convenience the " wave number " k instead of the wavelength itself. It is related to the wavelength by the equation $k = 2\pi/\lambda$. The symbol Δx represents the error in measurement of position. You can think of it as the standard deviation of our measurements. Similarly, Δk represents the error in measurement of k, or the wavelength.

Now, if we accept the de Broglie relationship, which states that the momentum of any particle is related to its wavelength by the simple formula $p = h/\lambda$, we find after a little manipulation of the deltas:

$$\Delta x \Delta p \sim h/2\pi.$$

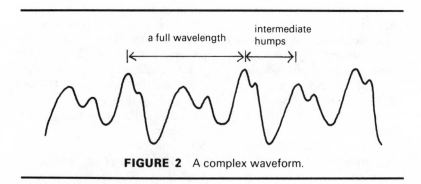

FIGURE 2 A complex waveform.

This is the famous Heisenberg uncertainty principle.* It tells us *there is a fundamental limit to the certainty with which we can measure both the position and momentum of a particle.* Notice that the only quantum mechanics involved was the introduction of the expression $p = h/\lambda$ for the momentum. Notice also, we have not said anything about how the position or momentum is measured. We have not mentioned Bohr's microscope, photons knocking electrons across the room, manna falling from heaven, or anything else. We can make this statement even stronger by noting that the classical result, $\Delta x \Delta k \sim 1$, is purely mathematical and absolutely does not depend on how the measurements were taken.

The Heisenberg uncertainty principle is one of the most important statements about nature. In the context of the present discussion, it is often invoked to explain why you cannot measure both the interference pattern *and* the slit through which the photon passed in the two-slit experiment. The explanation goes like this:

Instead of placing a counter behind the slits, use the setup in Fig. 3. The photographic plate is now mounted on springs so that any photon emerging from the top slit, like a bullet, will kick the screen downward. Just by recording the direction of the kick—that is, the particle's momentum—we can decide from which slit the photon emerged.† And since we retain the photographic plate to record a picture of the photons, we also retain the interference pattern. The paradox is resolved.

The paradox is not resolved. One can measure the kick given to the screen by the photon and thus know through which slit the photon passed. But this measurement of the momentum transfer requires that the screen's position be slightly uncertain at the start. That is, we do not know exactly where the screen was when the photon hit it. In fact, one cannot measure the kick given to the film without "jiggling the camera" just enough so that the picture is blurred, the interference pattern wiped out and the uncertainty principle always satisfied. Quantum theory again remains consistent.

We have spent a good deal of time discussing several phenomena that lie at the very heart of quantum mechanics. It is likely that if all of you have not been left totally confused, many of you will have at least been left troubled. If you are not troubled, then you should read the article again, because these phenomena have been troubling physi-

---

* A more exact derivation gives the result $\Delta x \Delta k \sim \frac{1}{2}$, and thus, $\Delta x \Delta p \sim h/4\pi$.

† For this argument, one must accept that photons, by and large, emerge as drawn. This can be arranged, if necessary, by putting a focusing lens to the right of the slits.

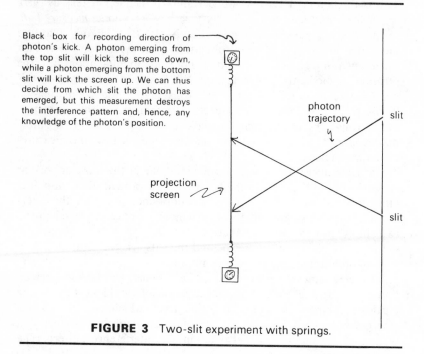

Black box for recording direction of photon's kick. A photon emerging from the top slit will kick the screen down, while a photon emerging from the bottom slit will kick the screen up. We can thus decide from which slit the photon has emerged, but this measurement destroys the interference pattern and, hence, any knowledge of the photon's position.

photon trajectory

slit

projection screen

slit

FIGURE 3   Two-slit experiment with springs.

cists for the last 60 years. Nonetheless, it is not just to disturb you that we have dwelt so long on interferometer slits and Perytons. There is a moral to our fable. It is a very simple moral.

Until this century, the realm of physics was very well defined. Physicists searched for eternal, immutable laws of nature. These laws of nature concerned inanimate objects. If man was to fit into the scheme of physics, then he fitted in only as a collection of inanimate objects, atoms and processes that involved these atoms. The laws that govern man's perception of his world did not qualify as physics. His erratic variability of senses was completely removed from the picture.

Now we find with quantum mechanics that we must put man back into the system and we find that the state of quantum mechanical systems is much more akin to the function of the human psyche. In the nineteenth century, no physicist alive would dare admit that the content of an observation, in any fundamental sense, depended upon the mode of perception. In the twentieth century, we have discovered that the content of our conscious minds is not independent of the mode of perception. In the two-slit experiment we could choose to measure either the photon's position or its momentum, but not both.

Only one was defined by our choice. Contrast this with the dissection of an automobile engine. While we cannot observe the carburetor before we open the air filter, we could, in principle, observe the entire engine simultaneously by taking an X-ray tomograph of the automobile. But with the two-slit interferometer, as with Perytons, a fundamental principle limits the observation to one quantity, depending on an act of our own will. ("Our will" does not necessarily refer to human will; a thinking robot could serve in the example, also.)

The attentive reader will have noticed a glaring gap in our discussion of the double-slit experiment. We never mentioned what *caused* the position of the photon to be recorded at a single point on the photographic screen. We claimed that the photon must have been spread out to pass through both slits and cause an interference pattern. We claimed the photon's position was registered on a single grain of film. You may well ask what is the cause that forced this extended object to "collapse" to a single point in order to darken one grain. This is much like asking for the source of a person's witty remark. Psychologists want to find hidden causes and reduce the domain of the psyche to that of physics. Freud's theory of wit, in which a clever remark is related to the release of a repressed thought, is considered a tremendous achievement. It is an effort to find a cause behind something that appears spontaneous.

Freud's theory appeared at a time when causes abounded in physics. The "collapse" of the wave packet is a phenomenon that appears uncaused. Physics is no longer the science Freud thought it was. How do we explain the "collapse" of the photon? We do not explain it. We conclude that either the explanation lies beyond physics or that physics encompasses more than was previously expected. As a great physicist, Eugene Wigner, was heard to say recently, "I think man is more than a collection of electrons." If wit and art are to be understood in terms of physics, then physics must change. Perhaps these uncaused processes, like the photon's collapse, will be the key to unlocking the indescribability of the human spirit.

In enlarging physics to include perception and awareness, we must steer clear of the simplistic world view of solipsism. However great one man's conviction that the world is flat, it does not mean that the world is flat. We propose a partial merging. Quantum mechanics may help resolve the Newtonian paradox mentioned at the beginning of this article: In order to make an observation on a system, you must interact with it and yet remain separate from it.

Looking back over the models discussed, those of you with practical minds will want to know, "How does viewing an ant colony as a wave, or a Peryton as a photon, help me pass my freshman physics course?" The question should be phrased as, "How does viewing a wave as an ant colony, or a photon as a Peryton, help me pass my freshman physics course?" We are sure there will be some good in it, visible in the long run if not in the short. You see, it is a common misconception that physicists are men who think in terms of memorized formulas and equations. To any physicist, an equation is not just a string of symbols, but a relationship. Good physicists think in terms of relationships that extend beyond equations. Great physicists see relationships among all things. An apple, a mirror, a cup of coffee, have produced theories of gravity, of relativity, of galaxies. There is no reason why the reverse should not apply. Let a knowledge of coherence affect the way you think about your relationships with those surrounding you. Let an understanding of the interference of waves and superposition allow you to decide when "two heads are better than one." We hope our collaboration has allowed a little insight into the way that physicists think. We hope the result was harmonious and coherent and that our individual intensities produced a whole greater than the sum of its parts. We hope that quantum mechanics will expand your world view and increase the intensity of the song in your heart. Let coherence be with you!

References

1. Melissus of Samos, as quoted in: Kirk & Raven, *The Presocratic Philosophers* (Cambridge University Press, Cambridge, 1971), pp. 304–305.
2. Lewis Thomas, *The Lives of a Cell* (Bantam Books, New York, 1975), pp. 12–13.
3. Jorge Luis Borges, *The Book of Imaginary Beings* (Avon Books, New York, 1970), pp. 182–183.

7

Extraterrestrial Intelligent Beings Do Not Exist

by Frank J. Tipler

Frank Tipler received his PhD at the University of Maryland in 1976 under Dieter Brill. His work is highly mathematical and involves the study of singularities—that is, the study of those regions of spacetime where most quantities one wishes to study become infinite. He is the author of the "no return" theorem, which shows that any state of the universe is unique and no return to a previous state is possible. As may have been gathered from "Grand Illusions," Frank has accrued a somewhat controversial reputation. In this article we finally allow him to speak for himself.

In this article it is argued that if extraterrestrial intelligent beings exist, then their spaceships must already be present in our solar system. Portions of this paper have appeared in three issues of The Quarterly Journal of the Royal Astronomical Society *(21, 267 [1980]; 22, 133 [1981]; 22, 279 [1981]), but this is the first printing of the full text. When originally published, the articles caused a minor furor and the author was besieged by newspaper and radio. In fairness to Frank, most of the criticisms leveled at his arguments were made by journalists who had clearly not read the paper. Whether one agrees with his position or not, the historical section of the article is, by any account, quite illuminating.*

Research for this article was supported by the National Science Foundation.

"Do there exist many worlds, or is there but a single world? This is one of the most noble and exalted questions in the study of nature."

ST. ALBERTUS MAGNUS (c. 1260)[1]*

"... the way whereby one can learn the pure truth concerning the plurality of worlds is by aerial navigation [space travel]."

PIERRE BOREL (1657)[1]

I. Introduction

One of the most interesting scientific questions is whether or not extraterrestrial intelligent beings exist. This question is not new; in one form or another it has been debated for thousands of years.† The contemporary advocates for the existence of such beings seem to be primarily astronomers and physicists (for example, Sagan,[2] Drake[3] and Morrison[4]), while most leading experts in evolutionary biology (for example, Dobzhansky,[5] Simpson,[6] Francois,[7] Ayala et al.[8] and Mayr[9]) contend that the earth is probably unique in harboring intelligence, at least among the planets of our galaxy. The biologists argue that the number of evolutionary pathways leading from one-celled organisms to intelligent beings is minuscule when compared with the total number of evolutionary pathways, and thus, even if we grant the existence of life on 10^9 to 10^{10} planets in our galaxy, the probability that intelligence has arisen in the Galaxy on any planet but our own is still very small. I agree with the biologists; I shall argue in this paper that the probability of the evolution of creatures with the technological capability of interstellar communication within 5 billion years after the development of life on an earthlike planet is less than 10^{-10}, and thus we are the only intelligent species now existing in this galaxy. The basic idea of my argument is straightforward, and indeed has led other authors (such as Fermi,[10] Dyson,[11] Hart,[12] Simpson[6]

---

* In the present article, superscript numbers of this type indicate the References, beginning on page 192, after the article and its three appendixes.

† See Appendix I for a brief history of the debate concerning the existence of extraterrestrial intelligent beings.

and Kuiper and Morris[13]) to conclude that extraterrestrial intelligent beings do not exist: if they did exist and possessed the technology for interstellar communication, they would also have developed interstellar travel and thus would already be present in our solar system. Since they are not here,[14,15] it follows that they do not exist. Although this argument has been expressed before, its force does not seem to have been appreciated. I shall try to rectify this situation by showing that an intelligent species with the technology for interstellar communication would necessarily develop the technology for interstellar travel, and this would automatically lead to the exploration and/or colonization of the Galaxy in less than 300 million years.

To begin with, we must assume that any intelligent species that develops the technology for interstellar communication must also have (or will develop in a few centuries) technology that is at least comparable to our present-day technology in other fields, particularly rocketry. This is actually a consequence of the principle of mediocrity[16] (that our own evolution is typical), which is usually invoked in analyses of interstellar communication. However, this assumption is also an essential one to make if interstellar communication via radio is to be regarded as likely. If we do not assume that an advanced species knows at least what we know, then we have no reason to believe an advanced species would use radio waves, for they may never have discovered such things. In the case of rocket technology, the human species developed rockets some 600 years before it was even aware of the existence of radio waves. Present-day chemical rockets can be regarded as logical extensions of early rocket technology.

In addition to a rocket technology comparable to our own, it seems likely that a species engaging in interstellar communication would possess a fairly sophisticated computer technology. In fact, Sagan has asserted:[17] "Communication with extraterrestrial intelligence . . . will require . . . , if our experience in radioastronomy is any guide, computer-actuated machines with abilities approaching what we might call intelligence." Furthermore, the Cyclops[18] and SETI[19] proposals for radiotelescopes to search for artificial extraterrestrial radio signals have required some fairly advanced data processing equipment. I shall therefore assume that any species engaging in interstellar communication will have a computer technology that is not only comparable to our present-day technology, but that is comparable to the level of technology that we know is possible, that we are now spending billions of dollars a year to develop and that a majority of computer experts believe we will actually possess within a century. That is, I shall assume that such a species will eventually develop a self-replicating universal constructor with intelligence comparable to the human level. Such a machine should be developed within a century, according to the experts.[20-22] This machine, combined with present-day rocket technology, would make it possible to explore and/or colonize the Galaxy in less than 300 million years, for an initial investment less than the cost of operating a 10-MW (megawatt) microwave beacon for several hundred years, as proposed in SETI.[19] It is a deficiency in computer technology, not rocket technology, that prevents us from beginning the exploration of the Galaxy tomorrow.

The conclusions of this paper may seem to hinge on the motivations of advanced extraterrestrial intelligent beings, a subject about which we admittedly know very little. However, the motivations of the most interesting class of intelligent beings, whose technology is far in advance of ours, are known *by definition*: they are *interested in communicating with us*, or otherwise interacting with us. It is this class that most SETI programs are designed to detect, and it is this class whose existence is conclusively ruled out by the arguments in this paper. I shall also argue that the interstellar exploration mechanism discussed in this paper has so many uses besides contacting other intelligent beings that *any* technologically advanced species would use it, and hence, if they existed, they would be here.

II. The General Theory of Space Exploration and Colonization

In space exploration (or colonization), one chooses a strategy that maximizes the probable rate of information gained (or regions colonized) and minimizes the cost of the information, subject to the constraints imposed by the level of technology. Costs may be minimized in two ways: first, "off-the-shelf" technology is to be used as far as possible to reduce the research and development costs; second, resources that could be used for no other purpose should be utilized as far as possible. The resources available in uninhabited stellar systems cannot be utilized for any human purpose unless a space vehicle is first sent; therefore, any optimal exploration strategy must utilize the material available in the other stellar system as far as possible. With present-day technology, such utilization could not be very extensive, but with the level of computer technology assumed in the previous section, these otherwise useless resources can be made to pay for virtually the entire cost of the exploration program.

What one needs is a self-reproducing universal constructor, which is a machine capable of making any device, given the construction materials and a construction program. In particular, it is capable of making a copy of itself. Von Neumann has shown[23,24] that such a machine is theoretically possible, and, in fact, a human being is a universal constructor specialized to perform on the surface of the earth. (Thus the manned space exploration [and colonization] program outlined in references 11–13 is just a special case of the exploration strategy discussed below.)

The payload of a probe to another stellar system would be a self-reproducing universal constructor with human-level intelligence (hereafter called a von Neumann machine), together with an engine for slowing down once the other stellar system is reached and an engine for traveling from one place to another within the target stellar system—the latter could be an electric propulsion system[25] or a solar sail.[26] This machine would be instructed to search out construction material with which to make several copies of itself and the original probe rocket engines. Judging from observations of our own solar system,[27] what observations we have of other stellar systems[28] and the vast majority of contemporary solar system formation theories,[29] such materials should be readily available in virtually any stellar system—including binary star systems—in the form of meteors, asteroids, comets and other debris from the formation of the stellar system. Whatever elements are necessary to reproduce the von Neumann machine, they should be available from one source or another. For example, the material in the asteroids is highly differentiated; many asteroids are largely nickel–iron, while others contain large amounts of hydrocarbons.[27]

As the copies of the space probe are made, they would be launched at the stars nearest the target star. When these probes reached these stars, the process would be repeated, and so on, until the probes had covered all the stars of the Galaxy. Once a sufficient number of copies had been made, the von Neumann machine would be programmed to explore the stellar system in which it finds itself and relay the information gained back to the original solar system from which the exploration began. In addition, the von Neumann machine could be programmed to use the resources of the stellar system to conduct scientific research that would be too expensive to conduct in the original solar system.

It would also be possible to use the von Neumann machine to colonize the stellar system. Even if there were no planets in the stellar system—the system could be a binary star system containing asteroidlike debris—the von Neumann machine could be programmed to turn some of this material into an O'Neill colony.[30] As to getting the inhabitants for the colony, it should be recalled that all the information needed to manufacture a human being is contained in the genes of a single human cell. Thus, if an extraterrestrial intelligent species possessed the knowledge to synthesize a living cell—and some experts[31,32] assert the human race could develop such knowledge within 30 years—they could program a von Neumann machine to synthesize a fertilized egg cell of their species. If they also possessed artificial womb technology—and such technology is in the beginning stages of being developed on earth[33]—then they could program the von Neumann machine to synthesize members of their species in the other stellar system. As suggested by Eiseley,[34] these beings could be raised to adulthood by robots in the O'Neill colony, after which they would be free to develop their own civilization in the other stellar system.

Suggestions have occasionally been made[35] that other solar systems could be colonized by sending frozen cells via space probe to the stars. However, it has not yet been shown[36–39] that such cells would remain viable over the long periods required to cross interstellar distances. This difficulty does not exist in the previously outlined colonization strategy; the computer memory of the von Neumann machine can be made so that it is essentially stable over long periods of time. If it is felt that the information required to synthesize an egg cell would tax the memory storage space of the original space probe, the information could be transmitted via microwave to the von Neumann machine once it has had time to construct additional storage capacity in the other solar system. The key point is that once a von Neumann machine has been sent to another solar system, the entire resources of that solar system become available to the intelligent species that controls the von Neumann machine; all sorts of otherwise too-expensive projects become possible. It would even be possible to program the von Neumann machine to construct a very powerful radio beacon with which to signal other intelligent species!

Thus the problem of interstellar travel has been reduced to the problem of transporting a von Neumann machine to another stellar system. This can be done even with present-day rocket technology. For example, Hunter[40,41] has pointed out that by using a Jupiter swingby to approach the sun and then giving a velocity boost at perihelion, a solar system escape velocity (v_{es}) of about 90 km/sec ($\sim 3 \times 10^{-4}c$; three ten-thousandths of the speed of light) is possible with present-day chemical rockets, even assuming the launch is made from the surface of the earth. As discussed in references 28 and 29, most other stars should have planets (or companion stars) with characteristics sufficiently close to those of the Jupiter–sun system to use this launch strategy in reverse to slow down in the other solar system.

The mass ratio μ (the ratio of the payload mass to the initial launch mass) for the initial acceleration would be 10^3, so the total trip would require $\mu < 10^6$ ("less than," since the 10^3 number assumed an earth-surface launch), quite high, but still feasible. (With Jupiter swingby only, the escape velocity would be $\sim 1.5 \times 10^{-4}c$ with $\mu = 10^3$.) The Voyager spacecraft will have[42] a solar escape velocity of about $0.6 \times 10^{-4}c$, with $\mu = 850$.

It thus seems reasonable to assume that any intelligent species would develop at least the rocket technology capable of a one-way trip with deceleration at the other stellar system, and with a travel velocity v_{es} of $3 \times 10^{-4}c$. At this velocity the travel time to the nearest stars would be between 10^4 and 10^5 years. This long travel time would necessitate a highly developed self-repair capacity, but this should be possible with the level of computer technology assumed for the payload.[43] Nuclear power sources could be developed that would supply power for the necessary length of time. However, nuclear power is not really necessary. If power utilization during the free-fall period was very low, even chemical reactions could be used to supply the power. As v_{es} is of the same order as the stellar random motion velocities, very sensitive guidance would be required, but this does not seem to be an insuperable problem with the assumed level of computer technology.

Because of the very long travel times, it is often argued that interstellar probes would be obsolete before they arrived.[44] However, in a fundamental sense, a von Neumann machine cannot become obsolete. The von Neumann machine can be instructed by radio to make the latest devices after it arrives at the destination star.

Restricting consideration to present-day rocket technology is probably too conservative. It is likely that an advanced intelligent species would eventually develop rocket technology at least to the limit that we regard as technically feasible today. For example, the nuclear pulse rocket of the Orion project pictured a solar escape velocity v_{es} of $3 \times 10^{-2}c$ with $\mu = 36$ for a one-way trip and deceleration at the target star.[45] The cost of the probe would be 3 trillion dollars in 1979 prices, almost all of the money being for the deuterium fuel. This is about the present GNP of the United States. Project Daedalus,[43] the interstellar probe study of the British Interplanetary Society, envisaged a stellar fly-by via nuclear pulse rocket (no slowdown at the target star), with $v_{es} = 1.6 \times 10^{-1}c$, $\mu = 150$, and a cost of 900 billion dollars. As before, almost all the cost is for the helium-3 fuel (at 1960 prices). With slowdown at the target star, $\mu = 2 \times 10^4$ and the cost would be 140 trillion dollars, or almost 100 times the United States GNP, and it would require centuries to extract the necessary helium-3 from the helium source in the Daedalus study, the Jovian atmosphere. The cost of such probes is far beyond present-day civilization. However, note that almost all the cost is for the rocket fuel. Building the probe itself and testing it would cost relatively little. A possible interstellar exploration strategy would be to design a probe capable of $v_{es} = 0.1c$, record the construction details in a von Neumann machine, launch the machine via a chemical rocket at $3 \times 10^{-4}c$ to a nearby stellar system and program the machine to construct and *fuel* several high-velocity ($0.1c$) probes with von Neumann payloads in the other system. When these probes reached their target stars, they would be programmed to build high-velocity probes, and so on. In this way the investment on interstellar probes by the intelligent species is reduced to a minimum while maximizing the rate at which the Galaxy is explored. (The von Neumann machines could conceivably be programmed to develop the necessary technology in the other system. This would reduce the initial investment even further.) The disadvantage, of course, is the fact that for 10^4 years there is no information on other stellar systems reaching the original solar system. There is a trade-off between the cost of the first probe and the time interval the intelligent species must wait before receiving any information on other stellar systems. But with second-generation probes with $v_{es} = 0.1c$, new solar systems could be explored at the rate of several per year by 10^5 years after the original launch. The intelligent species need only be patient and launch a sufficient number of initial probes at $v_{es} = 3 \times 10^{-4}c$ so that at least one succeeds in reproducing itself (or in making a high-velocity probe) several times. This number will depend on the failure rate. Project Daedalus[43] aimed at a mission failure rate of 10^{-4}, and the designers argued that such a failure rate was feasible with onboard repair. If we adopt this failure rate and assume failures to be statistically independent, then only three probes need be launched to reduce the failure probability to 10^{-12}. Judging by contemporary rocket technology, the cost of the initial low-velocity probes would be less than 10 billion dollars each, since von Neumann probes would be self-replicating and the original R&D costs would be small—von Neumann machines would originally be developed for other purposes.[46] Thus the exploration of the Galaxy would cost about 30 billion dollars—about the cost of the Apollo program.

To maximize the speed of exploration and/or colonization, one must minimize $[(d_{av}/v_{es}) + t_{const.}]$, where d_{av} is the average distance between stars and t_{const} is the time needed for the von Neumann machine to reproduce itself and the space probe. The time t_{const} will be much larger for $v_{es} = 0.1c$ probes than for $10^{-4}c$ probes. I would guess the minimum to be obtained for $v_{es} = 5 \times 10^{-2}c$ and $t_{const} =$

100 years. With d_{av} = 5 light-years (ly), this gives a rate of expansion of 2.5×10^{-2} ly/yr, and thus the Galaxy could be explored in 4 million years. For the purposes of this article, I shall assume only present-day rocket technology, which would give an expansion rate of 3×10^{-4} ly/yr, and the exploration of the Galaxy in 3×10^8 years.

This will be the expansion rate provided $t_{const} < 10^3$ yr. This seems a reasonable condition when we compare von Neumann probes with the only actual von Neumann machines of our experience, namely human beings. In their natural environment humans have a $t_{const} \sim$ 20–30 yr. If we compare a von Neumann probe to an entire technical civilization, then $t_{const} \sim 300$ yr for the time to build up the United States into an industrial nation. Most of this time was required to develop not the hardware but rather the knowledge of which machines to build. Possessing the necessary knowledge, Germany and Japan rebuilt their industries in a few decades after World War II, requiring only minor investment from outside. As for the t_{const} for space industries, O'Neill estimates[30] that space colonies could be self-sufficient and able to make more colonies in less than a century. Such a rapid space colony construction might require a large initial investment from earth, and this might correspond to a very large (i.e., expensive) probe payload. As before, the intelligent species could reduce the initial investment by building an initial probe small, but programmed to construct larger probes in the target systems. It seems unlikely that a Project Daedalus-size payload ($\sim 10^3$ tons), which seems to have most of the essential equipment of a von Neumann probe, would require longer than 100,000 years to reach the large-scale-probe-making stage. With this upper bound, the above estimate for the time needed to explore the Galaxy is valid. *Homo sapiens* has been in existence for less than 100,000 years.

Once the exploration and/or colonization of the Galaxy has begun, it can be modeled quite closely by the mathematical theory of island colonization—a theory developed fairly extensively by MacArthur and Wilson[47,48]—since the islands in the ocean are closely analogous to stars in the heaven and the von Neumann machines are even more closely analogous to biological species. There are several general conclusions applicable to interstellar exploration and/or colonization that follow from this theory. First, there are two basic behavioral strategies, the r-strategy and the K-strategy, which would be adopted in different phases of the colonization (r is the net reproductive rate [per capita births minus deaths] and K is the carrying capacity of the environment). The r-strategy is one of rapid reproduction at the expense of all else, and it would be followed in the early stages of the colonization. The K-strategy has a much smaller r; the emphasis is on securing the ecological niche in the target stellar system. The K-strategy would be adopted after the solar system had been colonized for some time and would result in fewer probes being sent to other stars. (In the past few centuries Western society has been an r-strategist, but as the carrying capacity of the environment is approached, it is beginning to adopt a K-strategy.) Second, the MacArthur–Wilson theory suggests[49] that the fraction of probes reaching a distance d from the system of launch could result in an exploration rate of up to twice the value of 2.5×10^{-2} ly/yr, with $v_{es} = 5 \times 10^{-2}c$ probes.

However, the MacArthur–Wilson theory must be modified before it can be applied to the problem of interstellar exploration/colonization. The MacArthur–Wilson theory assumes that the dispersal of colonizers is *random*, while the dispersal of von Neumann probes would not be. The von Neumann probes can use radio waves to determine which nearby stars have already been reached by other probes and launch descendant probes only at those stars which have not yet been reached. Animal colonizers do not have an analogous technique to learn of uninhabited but habitable islands, and so they must use a random search strategy. This also means that a diffusion model[50,51] of interstellar colonization would be incorrect. Diffusion is basically expansion against resistance, and there would be no resistance to the expansion of the volume of stars colonized by the von Neumann probes. In the case of diffusion of gas molecules, the diffusing molecules collide with molecules of the ambient gas. This leads (in the usual Brownian motion derivation of the one-dimensional diffusion equation) to an equally great probability of going backward as forward from a given collision site. Picture a one-dimensional array of points (stellar systems). The von Neumann probe at x_i would be programmed to send probes to all nearby unoccupied points (in the interval x_{i-r} to x_{i+r}, say) concentrating first on a probe to the x_{i+1} point. (It will have a memory of having arrived from the x_{i-j} point ($j \geq 1$), so the direction is defined.) If we neglect the reproductive failure rate of the probe at x_i, then with probability one the motion will be forward to x_{i+1}, x_{i+2}, etc., at a rate greater than or equal to $[(d_{av}/v_{es}) + t_{const}]$. By adjusting r (in effect, the net probe reproductive rate), we can cancel out the effect of the failure rate. Extending this analysis to three dimensions introduces no new question of principle. The expansion speed would still be $[(d_{av}/v_{es}) + t_{const}]$, at least in the later stages of expansion. (The earlier stages might be dominated by t_{const}, since there are more than two nearest neighbors. However, for t_{const} upper bounds like those given above, the time for expansion throughout the Galaxy would be dominated by the properties of its later stages.)

III. Astrophysical Constraints on the Evolution of Intelligent Species

The probability for the evolution in a star system of intelligent life that eventually attempts interstellar communication is usually expressed by the Drake equation:[52]

$$p = f_p n_e f_l f_i f_c$$

where f_p is the probability that a given star system will have planets, n_e is the number of habitable planets in a solar system that has planets, f_l is the probability that life evolves on a habitable planet, f_i is the probability that intelligent life evolves on a planet with life, and f_c is the probability that an intelligent species will attempt interstellar communication within 5 billion years after the formation of the planet on which it evolved. The time limit in f_c is tacit in most discussions of extraterrestrial intelligence. Some time period of approximately 5 billion years must be assumed in order to use the Drake equation to estimate the number of existing civilizations. If, for example, f_c were a Gaussian distribution with maximum at $t = 30$ billion years and σ (standard deviation) $= 1$ billion years, then we would be the only civilization in the Galaxy. The probability estimates made below will hold if it is assumed that f_c is either sharply peaked at 5 billion years after planetary formation or a Gaussian distribution with $t_{peak} < 6$ billion years and $\sigma > 1$ billion years.

The problem with the Drake equation is that only f_p—and to a lesser degree n_e—is subject to experimental determination. In order to measure a probability with a high degree of confidence, one must have a fairly large sample; for f_l, f_i and f_c, we have only one obvious case, the earth. However, if one accepts the argument that any intelligent species that attempts interstellar communication will begin the previously outlined galactic exploration program within 100 years after developing the technology for interstellar communication, then the sample size is enlarged to include all those stellar systems older than $t_{age} = 5$ billion years $+ t_{ex}$, where $t_{ex} \leq 300$ million years is the time needed to expand throughout the Galaxy. That is, the Drake probability p is less than or equal to $1/N$, where N is the number of stellar systems older than t_{age}, because all of these stars were, under the assumptions underlying the Drake equation, potential candidates to evolve communicating intelligent species; yet they failed to do so (had such species evolved on planets surrounding these stars within 5 billion years after star formation, their probes would already be present in the solar system, and these probes are not here).[14,15] Since f_p and n_e can in principle be determined by direct astrophysical measurement, the fact that extraterrestrial intelligent beings are not present in our solar system permits us to obtain a direct astrophysical measurement of an upper bound to the product $f_l f_i f_c$, which depends only on biological and sociological factors.

This argument assumes that the five probabilities of the Drake equation do not vary rapidly with galactic age. The available astrophysical evidence and most theories of the formation of solar systems indicate that this assumption is valid. The formation of solar systems requires that the interstellar gas be sufficiently enriched by "metals" (those elements heavier than helium). Most experts[29,53–55] agree that a substantial fraction of existing metals were formed in massive stars very early in galactic history—during the first 100 million years of the Galaxy's existence—and the metal abundance has changed by at most a factor of about two since then. The evidence[56,57] gives a galactic age between 11 and 18 billion years, and it is generally assumed[54] that the rate of star formation has been decreasing exponentially ever since the initial burst of heavy element formation. Existing stellar formation theory is unable to decide definitely if the so-called initial mass function—the number of stars formed per unit time with masses between m and $m + \Delta m$—changes with time after the initial burst of massive stars.[53] Furthermore, it is not clear to what extent the earthlike planet formation rate depends on the metal abundance.[58,59] However, the observational evidence[53] (such as it is) does not indicate a large variation of the initial mass function or the earthlike planet formation rate over time. I shall thus assume that these are roughly constant, and most discussions of extraterrestrial intelligence make the same assumption.[60,61] The factors f_l, f_i and f_c should not depend strongly on the evolution of the Galaxy as a whole (see, however, references 62–64), and so can be regarded as constants. Since the Galaxy is between 11 and 18 billion years old, the number N of stars older than 5.3 billion years is about twice the number of stars formed after the sun, and thus is approximately equal to the number of stars in the Galaxy, 10^{11}. Thus $p \leq 10^{-11}$. If we accept the usual values of $f_p = 0.1$ to 1, and $n_e = 1$, found in most discussions of interstellar communication,[2,18] then $f_l f_i f_c \leq 10^{-10}$. The number of communicating civilizations now existing in the Galaxy is less than or equal to $p \times$ (number of stars in Galaxy) = 1; that is to say, us.

This conclusion that we are the only technical civilization now existing in the Galaxy does not depend on any biological or sociological arguments except for the assumption that a communicating species would evolve in less than 5 billion years and would eventually begin interstellar travel; nor does it depend on f_p or n_e. It follows from just the interstellar travel assumption, the assumption that the galactic environment has not changed by more than a factor of five during the history of the Galaxy and the fact (?) that extraterrestrial probes are not present in our solar system.

If the galactic age is at the upper limit of 18×10^9 years or older, then we can conclude that we are the only technological species that now exists in the Galaxy around main-sequence stars of spectral type earlier than G3—assuming that such a species will develop interstellar travel before its star leaves the main sequence. (If the destruction of its solar system does not motivate a species to develop interstellar travel, it's hard to imagine what would.) Stars in this class will leave the main sequence in about 13×10^9 years or earlier, and so by the argument presented above, the number of species around such stars is less than $13.3/(18 - 13.3) \sim 3$. If we take into account the decrease of the star formation rate, we get 1 as the expected number of such species now existing.

For simplicity the above discussion was based on the Drake equation, but it should be clear that the same arguments can be used with any other plausible equation for the number of communicating species in the Galaxy—with the same results.

IV. Motivations for Interstellar Communication and Exploration

It is difficult to construct a plausible scenario whereby an intelligent species develops and retains for centuries an interest in interstellar communication together with the technology to engage in it, and yet does not attempt interstellar travel. Even if we adopt the pessimistic point of view that all intelligent species cease communication efforts before developing von Neumann machines—either because of a loss of interest or because they blow themselves to bits in a nuclear war—the conclusion that we are the only intelligent species in the Galaxy with interest in interstellar communication is not changed. For in this case, the longevity L of a communicating civilization is less than or equal to 100 years (using our computer experts' opinions for the time needed to develop von Neumann machines), and since the Drake equation gives $n = R_\star pL$ for the number of communicating civilizations in the Galaxy, we obtain $n = 10$, even if we use Sagan's optimistic estimate[2] of $R_\star p = 1/10$. (The number $R_\star$ is the average rate of star formation.) This value of n is essentially the same as $n \sim 1$ obtained in the previous section, and in any case, such short-lived civilizations would on the average be too far apart and exist for too short a time to engage in interstellar communication. (If $L \geq 100$ years, so that the species has time to develop probe technology, the value of L is irrelevant to the calculation of the number p. Once the probes have been launched, they will explore the Galaxy automatically; the death of the civilization that launched them would not stop them.) We are thus left with the possibility that for some reason, intelligent beings with the technology and desire for radio communication do not use the exploration strategy because they *choose* not to do so, not because they are incapable of developing the technology.

There is no good reason for believing this. Virtually any reason for engaging in interstellar radio communication provides an even stronger argument for the exploration of the Galaxy. For example, if the motivation for communication is to exchange information with another intelligent species, then as Bracewell[65,66] has pointed out, contact via space probe has several advantages over radio waves. One does not have to guess the frequency used by the other species, for instance. In fact, if the probe has a von Neumann machine payload, then the machine could construct an artifact in the solar system of the species to be contacted, an artifact so noticeable that it could not possibly be overlooked. If nothing else, the machine could construct a "Drink Coca-Cola" sign a thousand miles across and put it in orbit around the planet of the other species. Once the existence of the probe has been noted by the species to be contacted, information exchange can begin in a variety of ways. Using a von Neumann machine as a payload obviates the main objection[67] to interstellar probes as a method of contact, namely, the expense of putting a probe around each of an enormous number of stars. One need only construct a few probes, enough to make sure that at least one will succeed in making copies of itself in another solar system. Probes will then be sent to the other stars of the Galaxy automatically, with no further expense to the original species.

Morrison has expressed the opinion: "... once there is really interstellar communication, it may be followed by a ceremonial interstellar voyage of some special kind, which will not be taken for the sake of the information gained, or the chances for trade ..., but simply to be able to do it, for one special case, where there is a known destination. That's possible, one can imagine it being done—but it is very unlikely as a search procedure."[4] However, if it is granted that a *single* probe is launched, for *any* reason, then with a von Neumann machine payload, the same probe can be used to start the galactic expansion program outlined in section II. While enroute to a solar system known to be inhabited, the probe could make a stopover at a stellar system along the way, make several copies of itself, refuel and then proceed on its way (or send one of the copies to the inhabited system). If the inhabited system is further than 100 light-years, and if $v_{es} \leq 0.1c$ and $t_{const} \leq 100$ years, then the time needed to reach the inhabited system is increased by less than 10%, and one obtains the exploration and/or colonization of the entire Galaxy as a free bonus. Furthermore, because the inhabited system is so far away, *any* probe sent would have to be autonomous (which means that it would have to include a computer with a human-level intelligence) and be capable of self-repair (which means that it would essentially be a von Neumann machine). Since its instrumentation makes *any* interstellar probe capable of galactic exploration, why not use it for that purpose?

Consider the search strategy adopted by the *first* species arising in the Galaxy that was interested in interstellar communication. It would most likely be thousands or even millions of years before another such species arose. Even if another species arose simultaneously, the probability is only about 10^{-6} that it would be within 100 light-years of the other species. Therefore, when the first species begins to signal, it will probably get no answer for thousands or millions of years. During this time it will be receiving no information on other stellar systems for its investment. If strong interest in interstellar communication continues during this period, why should the species not also launch a few probes? *Some* information on other systems would be guaranteed in 100 to 10,000 years, even if other intelligent beings are not discovered. Also, if there are other intelligent beings in the Galaxy, the von Neumann probes will eventually find them, even if they are intelligent beings who would never develop on their own an interest in interstellar communication. If a null result is obtained from a radio wave search, there is always the possibility that the wrong frequency has been chosen or that some other means than radio waves has been used by the other species. There is no such problem with probes.

If human history is any guide, this first species will probably launch a probe rather than make radio beacons. In the early part of this century, when Lowell had convinced many that there were intelligent beings on Mars, but when interplanetary rocket probes were regarded as a ridiculous fantasy, the Harvard astronomer W. H. Pickering pointed out that communication with these beings was possible with a mirror one-half square mile in area: "[it] would be dazzlingly conspicuous to Martian observers, if they were intellectually and physically our equals."[68] If we were content to use such a device to learn about Mars from these hypothetical Martians, we would still know little about Mars. Instead, we sent probes, and Sagan's recent proposals[69] for advanced Mars probes are robots with manipulative ability and a considerable degree of artificial intelligence—they are a step in the direction of a von Neumann machine.

If we assume that a behavior pattern which is typical not only of *Homo sapiens* but also of all other living things on our planet would also be adopted by any intelligent species (to deny this would be to deny the assumption of mediocrity), then we would conclude that a sufficiently advanced intelligent species would launch von Neumann probes. All living things have a dispersal phase,[70] in which they tend to expand into new environments, for the dispersal behavior pattern is obviously selected by natural selection. The expansion is generally carried out to the limit imposed by genetic constitution. In intelligent species, this limit would be imposed by the level of technology,[71,72] and we would expect the dispersal behavior pattern to be present in at least some groups of an intelligent species. We should therefore expect that at least some groups of the species would attempt an expansion into the Galaxy, and the construction of only one successful von Neumann probe would be sufficient for this. By launching such a probe and using it to colonize the stars, a species increases the probability that it will survive the death of its star, nuclear war and other major catastrophes. Note that it need not take territory away from another species (intelligent or not) to accomplish this purpose. The species could, for example, restrict itself to the construction of O'Neill colonies around stars with no living things on their planets.

It is possible that an intelligent species that develops a level of technology capable of interstellar communication would decide not to build von Neumann machines because they would be afraid that they would lose control of the machines. Since no reproduction can be perfect, it is possible that the program that keeps the von Neumann probes under control of the intelligent species could accidentally be omitted during the reproduction process, with the result that the copy goes into business for itself. This problem can be avoided in three ways. First, the program that keeps the probe under control can be so integrated with the total program that its omission would cause the probe to fail to work at all. This is analogous to the constraints imposed on the cells used in recombinant DNA technology. Second, the intelligent species could program the probes to form colonies of the intelligent species in the stellar system reached by the probes. These colonies would be able to destroy any probes that slipped out of control. Third, the intelligent species might not care if the von Neumann machines slipped out of control. After all, a von Neumann machine would be an intelligent being in its own right, an intelligent being made of metal rather than flesh and blood. The rise of human civilization has been marked by a decline in racism and an extension of human rights—which include freedom—to a wider and wider class of people. If this trend continues and occurs in the cultures of all civilized beings, it seems likely that von Neumann machines would be recognized as fellow intelligent beings—beings that are the heirs to the civilization of the naturally evolved species that invented them, and with the right to the freedom possessed by the inventing species. If, on the other hand, the intelligent species retained their racism, it seems likely that they would regard other "flesh and blood" intelligent species as "nonpeople." If so, then they would either wish to avoid communication altogether (lest it "pollute" their culture with alien ideas), or else launch von Neumann machines to either colonize the Galaxy for themselves (lest it be done by "nonpeople" who would crowd them out) or to destroy these other intelligent species. For example, this colonization or destruction would be their best strategy if they believed that the biological "exclusion principle," which says[73,74] two species cannot occupy the same ecological niche in the same territory, applies to intelligent species. With the advent of the O'Neill colony, the ecological niche occupied by an intelligent species would consist of the entire material resources of a solar system. The ecological niches of two intelligent species would have to overlap. In any case, the von Neumann probes would be launched. If a species was not afraid of alien ideas itself, but was reluctant to contaminate the culture of another species with its own culture, then it should not attempt radio contact. However, with probes it would be possible to study an alien species without its becoming aware of the species that was studying it.

A final possibility to be considered is what I have hitherto denied, namely, that perhaps the von Neumann probes of an extraterrestrial intelligent species *are* present in our solar system. If a probe had just arrived, there would as yet be no evidence for its presence. The probability that a probe arrived for the first time within the past 20 years is 10^{-9} ($= 20/[\text{age of Galaxy}]$). Thus the probability that extraterrestrial intelligent beings exist but their probes have just arrived is actually greater than the calculated probability $f_l f_i f_c$ that they evolve. Another possibility would be that they are here but have decided for some reason not to make their presence known; this is the so-called zoo hypothesis.[75] Kuiper and Morris[13] have proposed testing this hypothesis by attempting to intercept radio communications between beings in our solar system and the parent stars. Another possible test would be to search for the construction activities of a von Neumann machine in our solar system. For example, one could look for the waste heat from such activities. As Dyson has pointed out,[11,76] this heat would give rise to an infrared excess, and the most likely place to look for a von Neumann probe would be the asteroid belt, where material is most readily available.[77] (It is amusing that most of the observed infrared radiation of astronomical origin does in fact come from the asteroid belt.[78]) If such a von Neumann probe were present in the solar system and if a large number of mutually intercommunicating intelligent species existed who were interested in studying us, we would expect the von Neumann machine to construct members of each of these species, together with spaceships, one appropriate type for each of the species. We would thus expect to see a wide variety of species and spaceships on earth studying us.* But no extraterrestrial ships of any type are seen.[14,15] Furthermore, if intelligent beings existed, it is likely that their probes would have arrived a billion years ago when there was nothing on earth but one-celled organisms; hence they would have no reason to hide their technology. The entire asteroid belt would be artifacts by now. Thus the evidence is enormous that extraterrestrial intelligent beings do not exist.

But the evidence is not utterly conclusive; beings with extremely advanced technology could be present in our solar system and make their presence undetectable should they wish to do so. The point is that a belief in the existence of extraterrestrial intelligent beings anywhere in the Galaxy is not significantly different from the widespread belief that UFOs are extraterrestrial spaceships. In fact, I strongly suspect the psychological motivation of both beliefs to be the same, namely, "The expectation that we are going to be saved from ourselves by some miraculous interstellar intervention. . . . "[79]

As discussed in reference 1, the belief in extraterrestrial intelligent beings is associated with a belief in the immensity of the cosmos: if there is a huge number of habitable planets, is it plausible that there is only one inhabited planet? I would contend the answer is yes. Wheeler has argued[80] that if the universe were much smaller than it is, it would terminate in a final singularity before intelligent life would have time to evolve. (This is an example of an "anthropic principle" argument. The anthropic principle[81,82,82a] states that many aspects of the universe are determined by the requirement that intelligent life exist in it.) Thus the universe must contain 10^{20} stars in order to contain a single intelligent species. We should not, therefore, be surprised if it indeed contains only one.

ACKNOWLEDGMENTS

I am grateful for extensive discussions on the subject of this paper with a large number of people, especially K. & P. Anderson, J. D. Barrow, R. Bracewell, J. H. Brooke, N. Calder, F. J. Dyson, R. O. Hansen, T. B. H. Kuiper, A. R. Martin, P. Morrison, C. Sagan, J. Silk, F. J. Tipler, Jr., J. A. Wheeler and P. Yasskin.

---

* Thus the argument by H. Y. Chiu in *Icarus*, **11**, 447 (1970) that UFOs could not be explained as spaceships because the observed visit rate would require too much material, is incorrect. Only one von Neumann probe need be sent to each solar system, and the material used for constructing the spaceships in each solar system could be reused.

A Brief History of the
Extraterrestrial Intelligence Concept

The idea that the heavenly bodies are inhabited is very old. In fact, the debate between believers in extraterrestrial intelligent beings and proponents of the earth as a uniquely inhabited world can be traced back to the ancient Greeks.[83] As will be seen in the following history, not only does the debate reoccur periodically as the centuries pass, but many of the arguments pro and con are reinvented as a new generation of debaters take up their pens.

Throughout history, the belief in a plurality of worlds is generally associated with three other beliefs. First, and most importantly, it is associated with what Lovejoy calls "the principle of plenitude," which asserts that what can exist must exist somewhere, and that if worlds like ours exist elsewhere in the universe, they must be inhabited by intelligent beings since no "genuine potentiality of being can remain unfulfilled."[84] The principle of plenitude has become the principle of mediocrity in twentieth-century discussions of extraterrestrial intelligence.

A second belief closely associated with the belief in extraterrestrial intelligence is the belief in the infinity of the cosmos, that there are an actual infinity of worlds. Although this belief is sometimes regarded as a consequence of the principle of plenitude, it does not, strictly speaking, follow from it since it is conceivable that there could be only a finite number of possible worlds. Nevertheless, people who argue for extraterrestrial intelligence have in past centuries generally argued for an infinite cosmos, while doubters in ETI have to either deny that the universe is infinite, or at most, admit that the universe is "indefinite" in extent.

Finally, believers in extraterrestrial intelligence have tended to lack what might be termed "a sense of history." A sense of history is more than an awareness that change, or even progressive change, has occurred in the past. It is also a feel for the contingent (at least in the eyes of human observers) nature of this change, an awareness of the uniqueness of events that are unpredictable because of their apparent insignificance at the time of occurrence, but whose effects amplify with time so as to exert an enormous, it not dominant, influence on future change. The role that the sense of history or the lack of it plays in the ETI debate will also be discussed in Appendix II.

Among the ancient Greeks and Romans, the term "world" signified what would now be called a Ptolemaic universe, consisting of a central earth, a single moon and sun, five planets and the fixed stars. A plurality of worlds meant, therefore, a number of self-contained universes, each with an inhabited central earth. Generally, supporters of a plurality of worlds in this sense also argued that the moon was of an earthly nature with intelligent inhabitants. The Pythagoreans; the atomists such as Democritus and Leucippus; the Stoics such as Epicurus and his follower Lucretius; Thales, Heraclitus and Plutarch all held both views in some form, and these are the most important supporters of the many-inhabited-worlds concept in antiquity.[83] The argument in favor of a plurality of inhabited worlds—especially those of Democritus and Epicurus—were based on the principle of plenitude and the idea that the universe is enormous. In the words of Metrodorus, a pupil of Epicurus, "It seems absurd that in a large field only one stalk should grow, and in an infinite space only one world exist."[85] There is no essential difference between this argument and present-day arguments for ETI based on the principle of mediocrity.

The most brilliant Greek thinkers were, however, opposed to the idea of a plurality of worlds. Plato, for instance, described those holding such a belief as possessing a sadly indefinite and ignorant mind.[86] He did admit that the question of the habitability of the planets was open, though he himself believed the earth was unique in this regard. The world system of Aristotle left no room either for a plurality of worlds or for inhabitants on the planets, and he argued at length against both these doctrines.[87] The plurality of worlds would be unstable because Aristotelian physics would require the earths of each world to come together at the center of the universe, and in any case, each finite world would have to be separated by a void, which is also an impossibility in his physics. The planets could not be inhabited because they were of a completely different substance than those found on earth. In short, the universe of Aristotle was *finite*, and the earth the only inhabited globe.

As is well known, the Aristotelian conception dominated thought until the time of Copernicus, and though the doctrine of a plurality of worlds (in the Greek sense) was occasionally discussed, it was rejected by most scholars, both pagan and Christian, in this period.[83] The Christian philosophers added two theological arguments against the plurality of worlds. Foremost was the idea of Christ's uniqueness (the uniqueness of the Incarnation): he appeared but *once* in the universe, and his appearance was a consequence of historical process peculiar to the earth. St. Augustine,[88] therefore, pointed out in the sixth century that if other intelligent beings similar to man existed, then they would also require a Saviour, which would contradict the uniqueness of Christ (1 Peter 3:18). Note that this argument is in part a historical argument. The uniqueness of the Incarnation is coupled with the notion of a unique historical or evolutionary process. This is the origin of the historical sense in Western thought.[89] Christianity was also somewhat anthropocentric—it tended to regard the universe as created for man—and this was a second theological argument against the plurality of worlds.

The great medieval scholars also rejected the idea of a plurality of worlds. During the scholastic flowering in the thirteenth century, St. Albertus Magnus[1] asserted, "Do there exist many worlds, or is there but a single world? This is one of the most noble and exalted questions in the study of nature." His discussions of this question were extensive[83] because he agreed with Augustine that a plurality of worlds would mean a plurality of Incarnations. In the end he too rejected the plurality, basing his argument on the uniqueness of Christ and Aristotelian physics. His reason for not rejecting the plurality idea out of hand was his belief in the principle of plenitude: if the power of God is infinite, why should he create merely a single finite world? By the principle of plenitude, this unlimited creative power should express itself by creating all possible worlds.

St. Thomas Aquinas, who was a pupil of Albertus Magnus and is regarded as the founder of scholastic philosophy, resolved this quandary by unqualifyingly rejecting both the plurality of worlds and the principle of plenitude.[90] He argued that if God made other worlds, they would either be similar or dissimilar to this one. If similar they would be in vain, which is not consistent with Divine Wisdom. If dissimilar, none would contain all things and therefore none would be perfect. An imperfect world could not be the work of a perfect Creator. Aquinas also rejected the idea of an infinite world.[91] He thus denies plurality, plenitude and infinity—the trinity of ideas that are associated throughout history. Despite the arguments of Aquinas, this trinity continued to be attractive to some medieval thinkers up to the time of Copernicus. For instance, the influential Saint Bonaventure contended that God could make a hundred worlds if he wished. He could suspend Aristotelian physics (i.e., the argument that two earths would come together) and create one in a place that is beyond the fixed stars.[82,92] (ETI believers have always been willing to suspend the physics of their day.[93]) Nicholas of Cusa, whose *De docta ignorantia* (1440) was the most influential book on cosmology until the seventeenth century,[93] was led by a belief in the principle of plenitude to accept the infinity and plurality of worlds. This work by Cusa had a considerable impact on the mystic Giordano Bruno,[94,95] who also advocated the infinity and plurality of worlds. With Bruno the notion of a plurality of worlds takes on its modern meaning—that of inhabited planets around a central sun—and "inhabited" signifies "inhabited by intelligent beings." Hereafter I shall use the phrase "plurality of worlds" to mean this, or to refer to inhabited planets in this solar system. Bruno is generally regarded as a martyr to science because he published the first defense of the Copernican system and was later burned at the stake. However, he defended Copernicus, not for scientific reasons, but for theological ones[93–100]: Indeed, he had contempt for the close mathematical reasoning of Copernicus[97,98,100] and was only interested in using Copernicus' work to attack some of the basic tenets of the Christian religion,[93,97] such as the uniqueness of the Incarnation. (His notion of time, like that of the ancient Stoic supporters of plurality, was cyclic.[93,97] He denied the Christian sense of history.) Bruno was executed for his mystical attack on Christianity, not for his belief in the plurality of worlds or his defense of Copernicus.[101] Far from being a martyr to science, Bruno actually harmed it, because the storm he raised caused the religious authorities to associate the Copernican system with anti-Christian agitation. This, in fact, was a major factor in the condemnation of Galileo.[101]

Kepler and Galileo, the two major figures in the early part of the Copernican revolution, were actually *opposed* to the idea of a plurality of inhabited worlds as put forward by Bruno. In his *Third Letter on Sunspots*, Galileo denounced "as false and damnable the view of those who would put inhabitants on Jupiter, Venus, Saturn, and the moon, meaning by 'inhabitants' animals like ours, and men in particular."[102,103] He claimed he could prove that other planets were without inhabitants of any kind.[102,103] Kepler, on the other hand, believed in the existence of living creatures on the planets, but he felt that they were definitely inferior to humans.[103] As to the question of inhabited planets around other stars, he regarded this as, at best, unsolved: "No moons have yet been seen revolving around [the stars]. Hence this will remain an open question until this phenomenon too is detected."[104] Kepler also argued against an infinite number of inhabited worlds in a manner similar to Aquinas: if there were infinitely many planets, there would be vain duplications; there would be "as many Galileos observing new stars in new worlds as there are worlds."[105] (The idea that an infinite universe implies an infinite number of identical people has recently [1978] been advanced by Ellis and Brundrit.[144])

In spite of the opposition of Kepler and Galileo, the plurality-of-worlds concept was given a very strong boost by the Copernican revolution. This occurred in several ways. First of all, the telescope disclosed mountains on the moon and satellites around Jupiter. These observations suggested that the planets were similar to the earth in gross structure. Second, the earth was demoted from the status of being an enormous body in the center of the universe to just one of six planets. To minds conditioned, because of the discovery of America in the previous century, to seeing unknown lands on this planet as inhabited places, it took but a small application of the principle of plenitude to envision the planets—regarded as distant lands—as also inhabited. Furthermore, the telescope had revealed innumerable stars, which were regarded as suns like our own. Teleological concepts such as the principle of plenitude were strong in the thinking of the day, and it was felt that the planets and stars must have been created for some purpose. Since the Copernican revolution had discredited the idea that these objects were created for our benefit—to light the night sky, for example—it was argued that they must have been created to be the abodes of other intelligent beings, just as the earth had been created for human beings.

The question of the plurality of worlds was often discussed in the seventeenth and eighteenth centuries, but rarely in scientific treatises. Rather, it was the subject matter of widely read books that were intended as popularizations of the new science. Probably the most influential of these was *Conversations on the Plurality of Worlds*, first published in 1686 by Bernard de Fontenelle, a novelist who later became secretary of the French Academy of Sciences. This work was a seventeenth-century best-seller. It went through numerous editions in French and was translated at least three times into English.[106] The book was written in the form of a series of conversations between Fontenelle and a lady, the Marchioness, who was ignorant of the new astronomy, but intelligent and anxious to learn. The arguments presented by Fontenelle in favor of a plurality of worlds were the teleological, the plenitude and the analogy (the earth and planets are, in the large, the same, hence they should be the same in having inhabitants) arguments outlined in the preceding paragraph. At the time Fontenelle was writing his *Conversations*, speculations on the possibility of interplanetary travel were widely discussed,[107] (e.g., by Pierre Borel[1,108]) and he embraced such ideas as his own:

... The art of flying is yet in its infancy, it may hereafter be brought to perfection; and the time may come when mankind may fly to the Moon.
I will not consent to this, said she [the Marchioness], that mankind

will ever carry the art of flying to such perfection, but that they will immediately break their necks. Very well, answered I [Fontenelle], if you will insist upon it, that mankind will always be such bad flyers, they may fly better in the Moon; the inhabitants of that Planet may be better formed for this trade than us; for it is very immaterial, whether we go there, or they come here, and we shall then be like the Americans, who could not form to themselves the idea of sailing so far on the sea, tho' at the other end of the world, they had long understood the art of Navigation.[109]

The Marchioness of 1686 immediately realized from this what many twentieth-century scientists have not realized, and what I have pointed out at length in the main body of this paper—namely, that if such intelligent beings on other planets existed and possessed interplanetary travel technology, they would have already arrived here on earth:

The people of the Moon would therefore have come to us before now, replied the Marchioness, almost in anger.[109]

There is but one reply to the Marchioness' objection if one still wishes to defend the plurality of worlds doctrine: one must argue that such beings would have had insufficient time since the creation of the universe to develop the necessary flight technology and fly to the earth. Fontenelle indeed makes such an answer:

The Europeans did not arrive in America till nearly at the end of six thousand years, replied I, breaking out into a laugh; this time was necessary for them to carry their Navigation to such perfection, so as to cross the Ocean. The people of the Moon know already, perhaps, how to make little flights in the Air; and at this time may be exercising themselves; when they shall be more able, and more experienced, we may see them; and God knows what will be our surprise.[110]

Six thousand years is indeed the approximate time the human race needed to develop from savagery to interplanetary travel. In Fontenelle's day it was also the age of the universe.[89] Nevertheless, the insufficient time scale argument was no more acceptable then than similar arguments are today. The Marchioness' opinion of the time scale argument was:

You are insupportable, said she, to push things so far with such weak reasoning.[110]

Fontenelle's book on the plurality of worlds was soon followed by two other very influential popular books supporting that belief. In 1698 the great Dutch scientist Christiaan Huygens, who discovered Saturn's rings and was instrumental in the formulation of the conservation of momentum law, published in Latin a defense of plurality entitled *The Celestial Worlds Discover'd.*[111] Widespread interest in plurality is indicated by the fact that an English translation appeared in the same year as the Latin version, and a French edition soon after. Huygens' book was followed in 1715 by William Derham's *Astro-Theology, or a Demonstration of the Being and Attributes of God from a Survey of the Heavens.*[112] Derham was a clergyman, the religious tutor of the Prince of Wales, and his book appeared under royal patronage.[113] Both of these books presented the Copernican system to the lay reader, and both argued for a plurality of inhabited worlds. The arguments in both books for a plurality were the usual ones mentioned above. Observations were cited to justify the rough similarity of the earth and the planets, and then the principle of plenitude was invoked to justify the existence of inhabitants. As Derham put it:

> Having thus represented the state of the Universe, according to the New [Copernican] System of it, the usual Question is, what is the use of so many Planets as we see about the Sun, and so many as are imagined to be about the Fixt Stars? To which the answer is, that they are *Worlds*, or places of habitation, which is concluded from their being habitable, and well provided for Habitation.[114]

(Note the lack of a sense of historical change in Derham's phrasing.) The existence of inhabited planets around each of the fixed stars was, as usual, justified by Derham on the basis of the principle of plenitude and antianthropocentric teleology:

> And if the Fixt Stars are so many Suns, certainly they minister to some grand uses in the Universe, far above what hath usually been attributed unto them. And what more probable uses, than to perform the office of so many suns? that is, to enlighten and warm as many Systems of Planets; after the manner as our Sun doth the Erraticks encompassing it. And that this is the Use and Office of the Fixt star is probable.
> 1. Because this is a far more probable and suitable use for so many Suns, so many glorious Bodies, than to say they were made only to enlighten and influence our lesser, and I may say inferior, Globe;
> 2. From the Parity, and constant Uniformity observable in all God's works, we have great reason to conclude that every Fixt Star hath a systeme of Planets, as well as the Sun.
> 3. This account of the Universe is far more magnificent, worthy of, and becoming the infinite CREATOR, than any other of the narrower schemes.[115]

Note the striking similarity between the wording of Derham's point 2 and Sagan's formulation of the principle of mediocrity.[2] A similar formulation is also to be found in Huygens' *Celestial Worlds*,[116] which Derham cites in defense of point 2.

In spite of the fact that the key arguments for a plurality of inhabited worlds were philosophical and theological rather than empirical—as Galileo pointed out,[117] there were no "sure observations" on the question of inhabitants, and the astronomer cannot affirm a thing to exist merely because it is logically possible—the belief in plurality was nearly unanimous in the world scientific community by the end of the eighteenth century. (I have been unable to discover any counterexamples.) This is probably due to the hold that the principle of plenitude had on men's minds, coupled with the fact that *all* popular works in astronomy at the time argued at length in favor of plurality and the scientific treatises mentioned plurality not at all, or only in passing. If there is no opposition to a view, it will become generally accepted whatever the evidence for it. (Witness the contemporary situation of ETI in the popular press.)

By the beginning of the nineteenth century, the belief in a plurality of worlds was regarded as so obviously true that plurality was used as an argument *against* the Incarnation, and hence against Christianity (the reverse of the situation from St. Augustine to Aquinas). As the American revolutionary Thomas Paine put it in his *Age of Reason* (1794):

> From whence then could arise the solitary and strange conceit, that the Almighty, who had millions of worlds equally dependent on his protection, should quit the care of all the rest, and come to die in our world, because they say one man and one woman had eaten an apple! And, on the other hand, are we to suppose that every world in the boundless creation had an Eve, an apple, a serpent, and redeemer? In this case, the person who is irreverently called the Son of God, and sometimes God himself, would have nothing else to do than to travel from world to world, in an endless succession of death, with scarcely a momentary interval of life.[118]

According to Thomas Chalmers, one of the most famous Scottish theologians of the early nineteenth century, this argument was a major source of intellectual doubts about Christianity in this period,[119] and he wrote an entire book, *Discourses on the Christian Revelation Viewed in Connection with the Modern Astronomy*, to refute it. This book was generally quoted in discussions on ETI during the following 40 years. What is remarkable is that in it Chalmers never once questions the existence of ETI. He can only argue that *Homo sapiens* may be the only intelligent species in the universe that fell from grace, an idea developed in the twentieth century by C. S. Lewis in his well-known science fiction trilogy.

The first person to take a critical *scientific* look at the empirical, as opposed to the philosophical or theological, evidence for extraterres-

trial intelligence was William Whewell, Master of Trinity College, Cambridge. Whewell pointed out that Chalmers' concessions were unnecessary, because all available evidence indicated that conditions on other planets of this solar system were so unlike those on earth as to render them uninhabitable by any form of life known to us. Furthermore, there was no evidence to show the existence of planets around the fixed stars analogous to the planets in our solar system. Whewell concluded that:

> The belief that other planets, as well as our own, are the seats of habitation of living things, has been entertained, in general, not in consequence of physical reasons, but in spite of physical reasons; and because there were conceived to be other reasons, of another kind, theological or philosophical, for such a belief.[120]

Whewell also attempted to refute some of these extrascientific reasons for a belief in a plurality of worlds. In particular, he argued *against* plenitude by emphasizing the *historical* change that the earth had undergone. By the geological evidence, humanity was very recent; for most of its history the earth was *uninhabited* by intelligent beings, and so Derham's argument for habitability loses its force. Whewell also doubted the infinity of the universe. However, the belief in ETI was by now too ingrained in popular culture for such arguments to make much headway. The publication of Whewell's book was greeted with a cry of outrage, almost as great as that which greeted Darwin's *Origin of Species* five years later. A great number of books and reviews appeared on the ETI question, most of them critical of Whewell's thesis.[121]

As is well known, there was a huge increase of interest in ETI in the latter part of the nineteenth century due to the claim that canals made by intelligent beings had been observed on Mars. These observations, popularized by the American astronomer Percival Lowell, were soon discounted by most professional astronomers,[122] but they retained a hold on the lay imagination for a century. Lowell's books about life on Mars provoked Alfred R. Wallace, who with Darwin was the discoverer of the theory of evolution by natural selection, into analyzing the likelihood of the evolution of an intelligent species elsewhere in the universe. He concluded that it was essentially zero, and thus we are alone in the universe. His arguments are worth repeating in detail, because although published in 1905, they are exactly the same as those given by modern evolutionists such as Dobzhansky, Simpson and Mayr. Thus, the biological arguments against the evolution of intelligence have not changed in 75 years. The great evolutionists have always been united against ETI. The biologists who have supported

ETI have generally been biologists with the viewpoint of a physicist, and lacking the historical sense of the evolutionist. Such men often err in questions about evolutionary biology; in particular, they err about questions concerning the probability of the evolution of a species with specified properties, as the recent recombinant DNA debate shows.[123]

Wallace's original arguments[124] against ETI were, like Whewell's, primarily physical, not biological. The evolutionary arguments against ETI appeared as an appendix to a later edition of his book, *Man's Place in Nature*:

> Those among my critics who have expressed adverse opinions, usually agree that my proofs of the absence of human life in the other planets of our system are very cogent if not quite conclusive, but declare that they cannot accept my view that the unknown planets that may exist around other suns are also without intelligent inhabitants. They give no reasons for this view other than the enormous number of suns that appear to be as favourably situated as our own, and the probability that many of them have planets as suitable as our earth for the development of human life. Several of them consider it absurd, or almost ludicrous, to suppose that man, or some being equally well organized and intelligent, has not been developed many times over in many of the worlds which they assume must exist. But not one of those who thus argue give any indication of having carefully weighed the evidence as to the number of very complex and antecedently improbable conditions which are absolutely essential for the development of the higher forms of organic life. . . .[125]

Wallace first pointed out that

> . . . if it is true that each species has arisen from one parental species, and one only [Wallace had earlier given evidence for this], then the whole line of descent from any living species (and therefore from man) back to the earliest form of life has been fixed and immutable; so that if any *one* of the thousand or millions of successive species in the line of descent had become extinct before it had been modified into the next species in the line of descent (or, which is the same thing, if it had been differently modified from what actually occurred), then that particular species which constitutes the last link in that particular line of descent—and this also applies to man—would never have come into existence.
>
> The ultimate development of man has, therefore, roughly speaking, depended on something like a million distinct modifications, each of a special type and dependent on some precedent changes in the organic and inorganic environments, or on both. The chances against such an enormously long series of definite modifications having occurred twice over, even in the same planet but in different isolated portions of it . . . are almost infinite, when we know how easily the balance of nature can be disturbed. . . .[126]

Thus, the existence of *Homo sapiens* on another planet was ruled out.

Wallace then generalized this argument from human beings to the class of intelligent beings:

> Of course, it may be said that a creature with a mind and spiritual nature equal to that of man might have been developed in a very diffterent form. . . . I briefly state why it seems quite inadmissible. In the first place, man differs from all other animals in the range and speciality of his mental nature even more than in his physical structure. It is generally admitted that his mental development has been rendered possible by a combination of three factors: the erect posture and free hand, the specialized vocal organization . . . and the exceptional development of the brain. . . .
>
> No other animal types make the slightest approach to any of these high faculties or show any indication of the possibility of their development. . . . The mere assertion, therefore, that a being possessing man's intellectual and moral nature combined with a very different animal form, might have been developed, is wholly valueless. We have no evidence for it, while the fact that no other animal than man *has* developed his special faculties even to a lower degree, is strong evidence against it . . . the evolutionary improbabilities now urged cannot be considered to be less than perhaps a hundred millions to one. . . .[127]

Wallace's arguments were ignored in future ETI debates; the modern evolutionists have independently rediscovered them (which argues for their correctness).

As Wallace mentioned in the above passage, the strongest argument for the existence of ETI *somewhere* in the universe hinged on there being an enormous number of planets. This assumption was generally accepted throughout the nineteenth century, for it seemed a consequence of the nebular hypothesis for the formation of the solar system, which was itself generally accepted.[93] However, this theory was rejected in the period 1900–1945 in favor of the "stellar collision theory,"[93] which made it appear that our solar system was a very rare phenomenon, if not unique. Consequently, most discussions of ETI in this period tended to reject the idea of ETI. This was certainly the case in the two most widely read popular books on astronomy in the period, *The Universe Around Us*[128] (1929) by Sir James Jeans and *The Nature of the Physical World*[129] (1928) by Sir Arthur Eddington. The latter in fact asserted: ". . . I feel inclined to claim that *at the present time* [Eddington's emphasis] our race is supreme. . . ."[129] Nevertheless, the principle of plenitude still exerted a strong hold on Eddington's mind, for he prefaced the above remark with: "I do not think that the whole of the Creation has been staked on the one planet where we live; and in the long run we cannot deem ourselves the only race that has been gifted with the mystery of consciousness." Nonscientists put more trust in plenitude than in scientific theories of

TWENTY-FOUR ART NOUVEAU POSTCARDS IN FULL COLOR FROM CLASSIC POSTERS, Hayward and Blanche Cirker. Ready-to-mail postcards reproduced from rare set of poster art. Works by Toulouse-Lautrec, Parrish, Steinlen, Mucha, Cheret, others. 12pp. 8¼× 11. 24389-3 Pa. $2.95

READY-TO-USE ART NOUVEAU BOOKMARKS IN FULL COLOR, Carol Belanger Grafton. 30 elegant bookmarks featuring graceful, flowing lines, foliate motifs, sensuous women characteristic of Art Nouveau. Perforated for easy detaching. 16pp. 8¼ × 11. 24305-2 Pa. $2.95

FRUIT KEY AND TWIG KEY TO TREES AND SHRUBS, William M. Harlow. Fruit key covers 120 deciduous and evergreen species; twig key covers 160 deciduous species. Easily used. Over 300 photographs. 126pp. 5⅜ × 8½. 20511-8 Pa. $2.25

LEONARDO DRAWINGS, Leonardo da Vinci. Plants, landscapes, human face and figure, etc., plus studies for Sforza monument, *Last Supper*, more. 60 illustrations. 64pp. 8¼ × 11⅛. 23951-9 Pa. $2.75

CLASSIC BASEBALL CARDS, edited by Bert R. Sugar. 98 classic cards on heavy stock, full color, perforated for detaching. Ruth, Cobb, Durocher, DiMaggio, H. Wagner, 99 others. Rare originals cost hundreds. 16pp. 8¼ × 11. 23498-3 Pa. $2.95

TREES OF THE EASTERN AND CENTRAL UNITED STATES AND CANADA, William M. Harlow. Best one-volume guide to 140 trees. Full descriptions, woodlore, range, etc. Over 600 illustrations. Handy size. 288pp. 4½ × 6⅜. 20395-6 Pa. $3.50

JUDY GARLAND PAPER DOLLS IN FULL COLOR, Tom Tierney. 3 Judy Garland paper dolls (teenager, grown-up, and mature woman) and 30 gorgeous costumes highlighting memorable career. Captions. 32pp. 9¼ × 12¼.
24404-0 Pa. $3.50

GREAT FASHION DESIGNS OF THE BELLE EPOQUE PAPER DOLLS IN FULL COLOR, Tom Tierney. Two dolls and 30 costumes meticulously rendered. Haute couture by Worth, Lanvin, Paquin, other greats late Victorian to WWI. 32pp. 9¼ × 12¼. 24425-3 Pa. $3.50

FASHION PAPER DOLLS FROM GODEY'S LADY'S BOOK, 1840-1854, Susan Johnston. In full color: 7 female fashion dolls with 50 costumes. Little girl's, bridal, riding, bathing, wedding, evening, everyday, etc. 32pp. 9¼ × 12¼.
23511-4 Pa. $3.50

THE BOOK OF THE SACRED MAGIC OF ABRAMELIN THE MAGE, translated by S. MacGregor Mathers. Medieval manuscript of ceremonial magic. Basic document in Aleister Crowley, Golden Dawn groups. 268pp. 5⅜ × 8½.
23211-5 Pa. $5.00

PETER RABBIT POSTCARDS IN FULL COLOR: 24 Ready-to-Mail Cards, Susan Whited LaBelle. Bunnies ice-skating, coloring Easter eggs, making valentines, many other charming scenes. 24 perforated full-color postcards, each measuring 4¼ × 6, on coated stock. 12pp. 9 × 12. 24617-5 Pa. $2.95

CELTIC HAND STROKE BY STROKE, A. Baker. Complete guide creating each letter of the alphabet in distinctive Celtic manner. Covers hand position, strokes, pens, inks, paper, more. Illustrated. 48pp. 8¼ × 11. 24336-2 Pa. $2.50

KEYBOARD WORKS FOR SOLO INSTRUMENTS, G.F. Handel. 35 neglected works from Handel's vast oeuvre, originally jotted down as improvisations. Includes Eight Great Suites, others. New sequence. 174pp. 9⅜ × 12¼.
24338-9 Pa. $7.50

AMERICAN LEAGUE BASEBALL CARD CLASSICS, Bert Randolph Sugar. 82 stars from 1900s to 60s on facsimile cards. Ruth, Cobb, Mantle, Williams, plus advertising, info, no duplications. Perforated, detachable. 16pp. 8¼ × 11.
24286-2 Pa. $2.95

A TREASURY OF CHARTED DESIGNS FOR NEEDLEWORKERS, Georgia Gorham and Jeanne Warth. 141 charted designs: owl, cat with yarn, tulips, piano, spinning wheel, covered bridge, Victorian house and many others. 48pp. 8¼ × 11.
23558-0 Pa. $1.95

DANISH FLORAL CHARTED DESIGNS, Gerda Bengtsson. Exquisite collection of over 40 different florals: anemone, Iceland poppy, wild fruit, pansies, many others. 45 illustrations. 48pp. 8¼ × 11. 23957-8 Pa. $1.75

OLD PHILADELPHIA IN EARLY PHOTOGRAPHS 1839-1914, Robert F. Looney. 215 photographs: panoramas, street scenes, landmarks, President-elect Lincoln's visit, 1876 Centennial Exposition, much more. 230pp. 8⅜ × 11¾.
23345-6 Pa. $9.95

PRELUDE TO MATHEMATICS, W.W. Sawyer. Noted mathematician's lively, stimulating account of non-Euclidean geometry, matrices, determinants, group theory, other topics. Emphasis on novel, striking aspects. 224pp. 5⅜ × 8½.
24401-6 Pa. $4.50

ADVENTURES WITH A MICROSCOPE, Richard Headstrom. 59 adventures with clothing fibers, protozoa, ferns and lichens, roots and leaves, much more. 142 illustrations. 232pp. 5⅜ × 8½. 23471-1 Pa. $3.50

IDENTIFYING ANIMAL TRACKS: MAMMALS, BIRDS, AND OTHER ANIMALS OF THE EASTERN UNITED STATES, Richard Headstrom. For hunters, naturalists, scouts, nature-lovers. Diagrams of tracks, tips on identification. 128pp. 5⅜ × 8. 24442-3 Pa. $3.50

VICTORIAN FASHIONS AND COSTUMES FROM HARPER'S BAZAR, 1867-1898, edited by Stella Blum. Day costumes, evening wear, sports clothes, shoes, hats, other accessories in over 1,000 detailed engravings. 320pp. 9⅜ × 12¼.
22990-4 Pa. $9.95

EVERYDAY FASHIONS OF THE TWENTIES AS PICTURED IN SEARS AND OTHER CATALOGS, edited by Stella Blum. Actual dress of the Roaring Twenties, with text by Stella Blum. Over 750 illustrations, captions. 156pp. 9 × 12.
24134-3 Pa. $7.95

HALL OF FAME BASEBALL CARDS, edited by Bert Randolph Sugar. Cy Young, Ted Williams, Lou Gehrig, and many other Hall of Fame greats on 92 full-color, detachable reprints of early baseball cards. No duplication of cards with *Classic Baseball Cards*. 16pp. 8¼ × 11. 23624-2 Pa. $2.95

THE ART OF HAND LETTERING, Helm Wotzkow. Course in hand lettering, Roman, Gothic, Italic, Block, Script. Tools, proportions, optical aspects, individual variation. Very quality conscious. Hundreds of specimens. 320pp. 5⅜ × 8½.
21797-3 Pa. $4.95

HOW THE OTHER HALF LIVES, Jacob A. Riis. Journalistic record of filth, degradation, upward drive in New York immigrant slums, shops, around 1900. New edition includes 100 original Riis photos, monuments of early photography. 233pp. 10 × 7⅞. 22012-5 Pa. $7.95

CHINA AND ITS PEOPLE IN EARLY PHOTOGRAPHS, John Thomson. In 200 black-and-white photographs of exceptional quality photographic pioneer Thomson captures the mountains, dwellings, monuments and people of 19th-century China. 272pp. 9⅜ × 12¼. 24393-1 Pa. $12.95

GODEY COSTUME PLATES IN COLOR FOR DECOUPAGE AND FRAMING, edited by Eleanor Hasbrouk Rawlings. 24 full-color engravings depicting 19th-century Parisian haute couture. Printed on one side only. 56pp. 8¼ × 11. 23879-2 Pa. $3.95

ART NOUVEAU STAINED GLASS PATTERN BOOK, Ed Sibbett, Jr. 104 projects using well-known themes of Art Nouveau: swirling forms, florals, peacocks, and sensuous women. 60pp. 8¼ × 11. 23577-7 Pa. $3.00

QUICK AND EASY PATCHWORK ON THE SEWING MACHINE: Susan Aylsworth Murwin and Suzzy Payne. Instructions, diagrams show exactly how to machine sew 12 quilts. 48pp. of templates. 50 figures. 80pp. 8¼ × 11. 23770-2 Pa. $3.50

THE STANDARD BOOK OF QUILT MAKING AND COLLECTING, Marguerite Ickis. Full information, full-sized patterns for making 46 traditional quilts, also 150 other patterns. 483 illustrations. 273pp. 6⅞ × 9⅝. 20582-7 Pa. $5.95

LETTERING AND ALPHABETS, J. Albert Cavanagh. 85 complete alphabets lettered in various styles; instructions for spacing, roughs, brushwork. 121pp. 8¾ × 8. 20053-1 Pa. $3.75

LETTER FORMS: 110 COMPLETE ALPHABETS, Frederick Lambert. 110 sets of capital letters; 16 lower case alphabets; 70 sets of numbers and other symbols. 110pp. 8⅛ × 11. 22872-X Pa. $4.50

ORCHIDS AS HOUSE PLANTS, Rebecca Tyson Northen. Grow cattleyas and many other kinds of orchids—in a window, in a case, or under artificial light. 63 illustrations. 148pp. 5⅜ × 8½. 23261-1 Pa. $2.95

THE MUSHROOM HANDBOOK, Louis C.C. Krieger. Still the best popular handbook. Full descriptions of 259 species, extremely thorough text, poisons, folklore, etc. 32 color plates; 126 other illustrations. 560pp. 5⅜ × 8½. 21861-9 Pa. $8.50

THE DORÉ BIBLE ILLUSTRATIONS, Gustave Doré. All wonderful, detailed plates: Adam and Eve, Flood, Babylon, life of Jesus, etc. Brief King James text with each plate. 241 plates. 241pp. 9 × 12. 23004-X Pa. $6.95

THE BOOK OF KELLS: Selected Plates in Full Color, edited by Blanche Cirker. 32 full-page plates from greatest manuscript-icon of early Middle Ages. Fantastic, mysterious. Publisher's Note. Captions. 32pp. 9⅜ × 12¼. 24345-1 Pa. $4.50

THE PERFECT WAGNERITE, George Bernard Shaw. Brilliant criticism of the Ring Cycle, with provocative interpretation of politics, economic theories behind the Ring. 136pp. 5⅜ × 8½. (Available in U.S. only) 21707-8 Pa. $3.00

CATALOG OF DOVER BOOKS

DECORATIVE NAPKIN FOLDING FOR BEGINNERS, Lillian Oppenheimer and Natalie Epstein. 22 different napkin folds in the shape of a heart, clown's hat, love knot, etc. 63 drawings. 48pp. 8¼ × 11. 23797-4 Pa. $1.95

DECORATIVE LABELS FOR HOME CANNING, PRESERVING, AND OTHER HOUSEHOLD AND GIFT USES, Theodore Menten. 128 gummed, perforated labels, beautifully printed in 2 colors. 12 versions. Adhere to metal, glass, wood, ceramics. 24pp. 8¼ × 11. 23219-0 Pa. $2.95

EARLY AMERICAN STENCILS ON WALLS AND FURNITURE, Janet Waring. Thorough coverage of 19th-century folk art: techniques, artifacts, surviving specimens. 166 illustrations, 7 in color. 147pp. of text. 7⅞ × 10¾. 21906-2 Pa. $8.95

AMERICAN ANTIQUE WEATHERVANES, A.B. & W.T. Westervelt. Extensively illustrated 1883 catalog exhibiting over 550 copper weathervanes and finials. Excellent primary source by one of the principal manufacturers. 104pp. 6⅛ × 9¼. 24396-6 Pa. $3.95

ART STUDENTS' ANATOMY, Edmond J. Farris. Long favorite in art schools. Basic elements, common positions, actions. Full text, 158 illustrations. 159pp. 5⅜ × 8½. 20744-7 Pa. $3.50

BRIDGMAN'S LIFE DRAWING, George B. Bridgman. More than 500 drawings and text teach you to abstract the body into its major masses. Also specific areas of anatomy. 192pp. 6½ × 9¼. (EA) 22710-3 Pa. $4.50

COMPLETE PRELUDES AND ETUDES FOR SOLO PIANO, Frederic Chopin. All 26 Preludes, all 27 Etudes by greatest composer of piano music. Authoritative Paderewski edition. 224pp. 9 × 12. (Available in U.S. only) 24052-5 Pa. $6.95

PIANO MUSIC 1888-1905, Claude Debussy. Deux Arabesques, Suite Bergamesque, Masques, 1st series of Images, etc. 9 others, in corrected editions. 175pp. 9⅜ × 12¼. (ECE) 22771-5 Pa. $5.95

TEDDY BEAR IRON-ON TRANSFER PATTERNS, Ted Menten. 80 iron-on transfer patterns of male and female Teddys in a wide variety of activities, poses, sizes. 48pp. 8¼ × 11. 24596-9 Pa. $2.00

A PICTURE HISTORY OF THE BROOKLYN BRIDGE, M.J. Shapiro. Profusely illustrated account of greatest engineering achievement of 19th century. 167 rare photos & engravings recall construction, human drama. Extensive, detailed text. 122pp. 8¼ × 11. 24403-2 Pa. $7.95

NEW YORK IN THE THIRTIES, Berenice Abbott. Noted photographer's fascinating study shows new buildings that have become famous and old sights that have disappeared forever. 97 photographs. 97pp. 11⅜ × 10. 22967-X Pa. $6.50

MATHEMATICAL TABLES AND FORMULAS, Robert D. Carmichael and Edwin R. Smith. Logarithms, sines, tangents, trig functions, powers, roots, reciprocals, exponential and hyperbolic functions, formulas and theorems. 269pp. 5⅜ × 8½. 60111-0 Pa. $3.75

HANDBOOK OF MATHEMATICAL FUNCTIONS WITH FORMULAS, GRAPHS, AND MATHEMATICAL TABLES, edited by Milton Abramowitz and Irene A. Stegun. Vast compendium: 29 sets of tables, some to as high as 20 places. 1,046pp. 8 × 10½. 61272-4 Pa. $19.95

REASON IN ART, George Santayana. Renowned philosopher's provocative, seminal treatment of basis of art in instinct and experience. Volume Four of *The Life of Reason*. 230pp. 5⅜ × 8.   24358-3 Pa. $4.50

LANGUAGE, TRUTH AND LOGIC, Alfred J. Ayer. Famous, clear introduction to Vienna, Cambridge schools of Logical Positivism. Role of philosophy, elimination of metaphysics, nature of analysis, etc. 160pp. 5⅜ × 8½. (USCO)
20010-8 Pa. $2.75

BASIC ELECTRONICS, U.S. Bureau of Naval Personnel. Electron tubes, circuits, antennas, AM, FM, and CW transmission and receiving, etc. 560 illustrations. 567pp. 6½ × 9¼.   21076-6 Pa. $8.95

THE ART DECO STYLE, edited by Theodore Menten. Furniture, jewelry, metalwork, ceramics, fabrics, lighting fixtures, interior decors, exteriors, graphics from pure French sources. Over 400 photographs. 183pp. 8⅜ × 11¼.
22824-X Pa. $6.95

THE FOUR BOOKS OF ARCHITECTURE, Andrea Palladio. 16th-century classic covers classical architectural remains, Renaissance revivals, classical orders, etc. 1738 Ware English edition. 216 plates. 110pp. of text. 9½ × 12¾.
21308-0 Pa. $10.00

THE WIT AND HUMOR OF OSCAR WILDE, edited by Alvin Redman. More than 1000 ripostes, paradoxes, wisecracks: Work is the curse of the drinking classes, I can resist everything except temptations, etc. 258pp. 5⅜ × 8½. (USCO)
20602-5 Pa. $3.50

THE DEVIL'S DICTIONARY, Ambrose Bierce. Barbed, bitter, brilliant witticisms in the form of a dictionary. Best, most ferocious satire America has produced. 145pp. 5⅜ × 8½.   20487-1 Pa. $2.50

ERTÉ'S FASHION DESIGNS, Erté. 210 black-and-white inventions from *Harper's Bazar*, 1918-32, plus 8pp. full-color covers. Captions. 88pp. 9 × 12.
24203-X Pa. $6.50

ERTÉ GRAPHICS, Erté. Collection of striking color graphics: *Seasons, Alphabet, Numerals, Aces* and *Precious Stones*. 50 plates, including 4 on covers. 48pp. 9⅜ × 12¼.   23580-7 Pa. $6.95

PAPER FOLDING FOR BEGINNERS, William D. Murray and Francis J. Rigney. Clearest book for making origami sail boats, roosters, frogs that move legs, etc. 40 projects. More than 275 illustrations. 94pp. 5⅜ × 8½.   20713-7 Pa. $1.95

ORIGAMI FOR THE ENTHUSIAST, John Montroll. Fish, ostrich, peacock, squirrel, rhinoceros, Pegasus, 19 other intricate subjects. Instructions. Diagrams. 128pp. 9 × 12.   23799-0 Pa. $4.95

CROCHETING NOVELTY POT HOLDERS, edited by Linda Macho. 64 useful, whimsical pot holders feature kitchen themes, animals, flowers, other novelties. Surprisingly easy to crochet. Complete instructions. 48pp. 8¼ × 11.
24296-X Pa. $1.95

CROCHETING DOILIES, edited by Rita Weiss. Irish Crochet, Jewel, Star Wheel, Vanity Fair and more. Also luncheon and console sets, runners and centerpieces. 51 illustrations. 48pp. 8¼ × 11.   23424-X Pa. $2.00

YUCATAN BEFORE AND AFTER THE CONQUEST, Diego de Landa. Only significant account of Yucatan written in the early post-Conquest era. Translated by William Gates. Over 120 illustrations. 162pp. 5⅜ × 8½. 23622-6 Pa. $3.50

ORNATE PICTORIAL CALLIGRAPHY, E.A. Lupfer. Complete instructions, over 150 examples help you create magnificent "flourishes" from which beautiful animals and objects gracefully emerge. 8⅛ × 11. 21957-7 Pa. $2.95

DOLLY DINGLE PAPER DOLLS, Grace Drayton. Cute chubby children by same artist who did Campbell Kids. Rare plates from 1910s. 30 paper dolls and over 100 outfits reproduced in full color. 32pp. 9¼ × 12¼. 23711-7 Pa. $2.95

CURIOUS GEORGE PAPER DOLLS IN FULL COLOR, H. A. Rey, Kathy Allert. Naughty little monkey-hero of children's books in two doll figures, plus 48 full-color costumes: pirate, Indian chief, fireman, more. 32pp. 9¼ × 12¼. 24386-9 Pa. $3.50

GERMAN: HOW TO SPEAK AND WRITE IT, Joseph Rosenberg. Like *French, How to Speak and Write It.* Very rich modern course, with a wealth of pictorial material. 330 illustrations. 384pp. 5⅜ × 8½. (USUKO) 20271-2 Pa. $4.75

CATS AND KITTENS: 24 Ready-to-Mail Color Photo Postcards, D. Holby. Handsome collection; feline in a variety of adorable poses. Identifications. 12pp. on postcard stock. 8¼ × 11. 24469-5 Pa. $2.95

MARILYN MONROE PAPER DOLLS, Tom Tierney. 31 full-color designs on heavy stock, from *The Asphalt Jungle, Gentlemen Prefer Blondes,* 22 others. 1 doll. 16 plates. 32pp. 9⅜ × 12¼. 23769-9 Pa. $3.50

FUNDAMENTALS OF LAYOUT, F.H. Wills. All phases of layout design discussed and illustrated in 121 illustrations. Indispensable as student's text or handbook for professional. 124pp. 8⅛ × 11. 21279-3 Pa. $4.50

FANTASTIC SUPER STICKERS, Ed Sibbett, Jr. 75 colorful pressure-sensitive stickers. Peel off and place for a touch of pizzazz: clowns, penguins, teddy bears, etc. Full color. 16pp. 8¼ × 11. 24471-7 Pa. $2.95

LABELS FOR ALL OCCASIONS, Ed Sibbett, Jr. 6 labels each of 16 different designs—baroque, art nouveau, art deco, Pennsylvania Dutch, etc.—in full color. 24pp. 8¼ × 11. 23688-9 Pa. $2.95

HOW TO CALCULATE QUICKLY: RAPID METHODS IN BASIC MATHE-MATICS, Henry Sticker. Addition, subtraction, multiplication, division, checks, etc. More than 8000 problems, solutions. 185pp. 5 × 7¼. 20295-X Pa. $2.95

THE CAT COLORING BOOK, Karen Baldauski. Handsome, realistic renderings of 40 splendid felines, from American shorthair to exotic types. 44 plates. Captions. 48pp. 8¼ × 11. 24011-8 Pa. $2.25

THE TALE OF PETER RABBIT, Beatrix Potter. The inimitable Peter's terrifying adventure in Mr. McGregor's garden, with all 27 wonderful, full-color Potter illustrations. 55pp. 4¼ × 5½. (Available in U.S. only) 22827-4 Pa. $1.50

BASIC ELECTRICITY, U.S. Bureau of Naval Personnel. Batteries, circuits, conductors, AC and DC, inductance and capacitance, generators, motors, trans-formers, amplifiers, etc. 349 illustrations. 448pp. 6½ × 9¼. 20973-3 Pa. $7.95

CATALOG OF DOVER BOOKS

THE PRINCIPLE OF RELATIVITY, Albert Einstein et al. Eleven most important original papers on special and general theories. Seven by Einstein, two by Lorentz, one each by Minkowski and Weyl. 216pp. 5⅜ × 8½.   60081-5 Pa. $3.50

PINEAPPLE CROCHET DESIGNS, edited by Rita Weiss. The most popular crochet design. Choose from doilies, luncheon sets, bedspreads, apron—34 in all. 32 photographs. 48pp. 8¼ × 11.   23939-X Pa. $2.00

REPEATS AND BORDERS IRON-ON TRANSFER PATTERNS, edited by Rita Weiss. Lovely florals, geometrics, fruits, animals, Art Nouveau, Art Deco and more. 48pp. 8¼ × 11.   23428-2 Pa. $1.95

SCIENCE-FICTION AND HORROR MOVIE POSTERS IN FULL COLOR, edited by Alan Adler. Large, full-color posters for 46 films including *King Kong, Godzilla, The Illustrated Man*, and more. A bug-eyed bonanza of scantily clad women, monsters and assorted other creatures. 48pp. 10¼ × 14¼.   23452-5 Pa. $8.95

TECHNICAL MANUAL AND DICTIONARY OF CLASSICAL BALLET, Gail Grant. Defines, explains, comments on steps, movements, poses and concepts. 15-page pictorial section. Basic book for student, viewer. 127pp. 5⅜ × 8½.
     21843-0 Pa. $2.95

STORYBOOK MAZES, Dave Phillips. 23 stories and mazes on two-page spreads: *Wizard of Oz, Treasure Island, Robin Hood*, etc. Solutions. 64pp. 8¼ × 11.
     23628-5 Pa. $2.25

PUNCH-OUT PUZZLE KIT, K. Fulves. Engaging, self-contained space age entertainments. Ready-to-use pieces, diagrams, detailed solutions. Challenge a robot, split the atom, more. 40pp. 8¼ × 11.   24307-9 Pa. $3.50

THE HUMAN FIGURE IN MOTION, Eadweard Muybridge. Over 4500 19th-century photos showing stopped-action sequences of undraped men, women, children jumping, running, sitting, other actions. Monumental collection. 390pp. 7⅞ × 10⅝.   20204-6 Clothbd. $18.95

PHOTOGRAPHIC SKETCHBOOK OF THE CIVIL WAR, Alexander Gardner. Reproduction of 1866 volume with 100 on-the-field photographs: Manassas, Lincoln on battlefield, slave pens, etc. 224pp. 10⅝ × 8¼.   22731-6 Pa. $6.95

FLORAL IRON-ON TRANSFER PATTERNS, edited by Rita Weiss. 55 floral designs, large and small, realistic, stylized; poppies, iris, roses, etc. Victorian, modern. Instructions. 48pp. 8¼ × 11.   23248-4 Pa. $1.95

AUTOBIOGRAPHY: The Story of My Experiments with Truth, Mohandas K. Gandhi. Boyhood, legal studies, purification, the growth of the Satyagraha (nonviolent protest) movement. Critical, inspiring work of the man who freed India. 480pp. 5⅜ × 8½.   24593-4 Pa. $6.95

ON THE IMPROVEMENT OF THE UNDERSTANDING, Benedict Spinoza. Also contains *Ethics, Correspondence*, all in excellent R Elwes translation. Basic works on entry to philosophy, pantheism, exchange of ideas with great contemporaries. 420pp. 5⅜ × 8½.   20250-X Pa. $5.95

Prices subject to change without notice.

Available at your book dealer or write for free catalog to Dept. GI, Dover Publications, Inc., 31 East 2nd St. Mineola, N.Y. 11501. Dover publishes more than 175 books each year on science, elementary and advanced mathematics, biology, music, art, literary history, social sciences and other areas.